"I have been a rubber tapper, I am a rubber tapper. I lived in the muddy swamps, in the loneliness of the forests, with my crew of malaria-ridden men, cutting the bark of trees that have white blood, like that of the gods."

J.E. Rivera
La Vortex
1935

John C. Yungjohann

White Gold

The diary of a rubber cutter
in the Amazon
1906 - 1916

by

John C. Yungjohann

Edited by

Ghillean T. Prance

Epilogue by
Yungjohann Hillman

Photographs by
Ghillean T. Prance

SYNERGETIC PRESS

Published by Synergetic Press, Inc.
P.O. Box 689
Oracle, Arizona 85623

Cover illustration by Percy Lau.
Cover and book design by Kathleen Dyhr.
Typesetting by Synergetic Press.

ISBN 0 907791 16 6

Printed in the United States of America by Arizona Lithographers, Tucson.

Introduction

This is a book about survival. The remarkable story of a rubber cutter who survived through the most incredible adventures, sickness and exploitation in the upper reaches of the Amazon at the turn of the century in the height of the Amazon rubber boom. The hero, John C. Yungjohann kept a diary and wrote a personal account of what it meant to be an Amazon rubber cutter. It is little realized that many North Americans and Europeans joined the Brazilians and the enslaved Indians as abused and exploited workers for the Amazon rubber barons. This first-hand account written in English, shows how the debt system worked and how difficult it was for anyone to get out of the burden of his debt. It recounts how all the writer's friends gradually died off and he was frequently left alone. It is a story of human determination and how sheer will power can pull one through the most difficult circumstances. Before our writer tells his incredible story a few words of explanation must be given.

The diary came to my hands as a contemporary Amazon explorer because I happened to meet the grandson of the writer when he was, appropriately, in the middle of the Amazon jungle on a pilgrimage to the haunts of his grandfather. Yungjohann Hillman decided to go the Amazon in 1982 to try and experience some of the atmosphere of his grandfather's adventures, and so we met in Iquitos, Peru. After travelling with him in a luxury liner to Amatura in Brazil I left him in that small riverside town to make his own way down the Amazon river to Manaus on local river launches. I too had become fascinated by the adventures of his grandfather, and have been privileged to go through his writings to edit and explain them. As can be seen from some of my illustrations, all from the contemporary Amazon scene, some things have not changed too much, others fortunately have. There is much less exploitation of people in the region although riverside trading launches still over-charge for their goods and buy rubber and other commodities such as hearts of palm at extremely low prices.

It is appropriate to bring out this diary from the turn of the century at a time when today's rubber gatherers are involved in a new struggle — the right to keep some forest standing so that they can continue their work amid the rampant deforestation that is occurring in Amazonia. In September 1987, Amazonian rubber gatherer Francisco Mendes Filho, President of the rubber gatherers union was awarded one of the Better World Medals from the Better World Society, for his work to protect the forest of Amazonia. This man, together with many fellow rubber gatherers, is lobbying hard for the establishment of extraction forests where they can continue to extract rubber, Brazil nuts and other forest products. This is obviously a much better use of the forest than for cattle pasture. Things have not changed much since the time of Yungjohann. Mendes has so far survived four assassination attempts on account of his challenge to the contemporary economic system. Chico Mendes is a worthy successor to John Yungjohann.[1]

Another unfortunate change today is the reduction in the number of wild animals. As I have travelled on the rivers of Acre, alligators or caiman are rare, the turtles few and far between, one hardly ever sees a boa constrictor and the jaguars and pumas have been practically eliminated by hunters after their skins. In today's Amazon one is therefore, unfortunately, unlikely to have some of the animal adventures encountered by Yungjohann. Many of the early Amazon explorers recount this abundance of wildlife. It is a sad comment that the western occupation of the region has led to so much extermination of game. The indigenous population, which was quite large, never put the same pressure on the animals.

In editing this account I have left the writer's own expressions rather than try to produce it all in grammatically correct English. The changes which I have made are only where the sense is not clear or to correct minor factual errors, such as the spelling of various Portuguese words. This story is all the more interesting maintained in the original colloquial style of our rubber gatherer hero. Points which are unclear or need some explanations are clarified in a series of footnotes. A few terms which occur repeatedly and are key to understanding the region are defined in the glossary at the end to avoid repeated footnotes. The chapter divisions are those of the author not the editor. For readers who are unfamiliar with the Amazon rubber boom, a short bibliography of some of the best reading on the subject is given at the end of this book.

---

[1] Mr. Mendes was tragically murdered on the 22nd of December 1988 for his work in defending the rights of rubber cutters to keep uncut rainforest for their livelihood.

I thank Yungjohann Hillman for his collaboration with this work and for sharing so readily his grandfather's account. I hope that this will help to document what some of the horrors and the daily life of the rubber cutter were like at the height of the Amazon rubber boom. It should be of historic interest to all who are interested in the story of the economically important tree *Hevea brasiliensis,* the Amazon rubber. It tells us a lot of first-hand facts about rubber collecting and so is an important contribution to the science of economic botany. However, it also reads like a soap opera and is therefore of a much wider general interest.

Travel in Amazonia today is tame compared to what you will read here. However, even in the day of air service, new highways and electronic communications the contemporary explorer can experience a little of the atmosphere recounted here. Perhaps what I most admire is that our hero, John Yungjohann was carried through all his adventures by his strong will to survive. With this spirit the human body can undergo remarkable devastation and still survive. My one experience of this was the time I got malaria on another tributary of the Rio Purus, the Ituxí River. We had arrived at our destination, the end of the small Curuquetê river where our inflatable boats could not penetrate the overhanging vegetation any further. Our plan was to cross on foot from the Purus watershed to the Madeira watershed, but the day we reached the river's end without gasoline for return was the day the fevers and chills began. After discussion with fellow expedition members it was decided that the only thing to do was for me to walk out the three day hike over the watershed with one of our field guides. We set out the next day carrying only light rucksacks with a hammock and food. The fever of malaria comes in phases and when the fever came on I put up my hammock and rested, and as soon as the fever went down I walked. To make matters worse, on the first day our guide lost his way to the one house and trail over the watershed. Three times we found ourselves back at the same grove of Brazil nut trees at a time that every step was an effort and a strain and was using up what little strength I had left in me. At dusk we finally reached the house where we spent the night, the five hour walk had taken us fourteen hours. By morning my guide was also very sick with malaria. He appeared to be in far worse shape than I. We sent a young boy from the house on ahead of us to get a mule from the edge of the Rio Madeira for my use for the last part of the hike. We set out again slowly and forcing ourselves to walk. I remember every step telling myself you must go on. Rest stops tended to be longer and I was unable to hold down the bananas we were given at the house. The third day the mule came to us and by now my companion was unable to walk. With great effort we pushed him up on the mule which was sent for me, and I proceeded along the trail leading a rather stubborn mule.

I remember lying down on the middle of the trail and saying, "This is it, I can't go on." After a few minutes thinking, for fortunately my mind was still clear, I thought that: "You can either stay here and die or you just have to get up and walk." I have never before or again forced my unwilling body in such a way, but somehow I managed to combine putting one foot in front of the other, although I began to stumble more frequently when we came to obstacles on the trail.

Perhaps the worst time was when we forded across a stream. The cool water felt good on my fevered body as I waded up to my waist. However, the mule stumbled and threw my companion off into the stream. I then had to grab this half unconscious body and rescue him. I will never know how in my weakened state I lifted him back on the mule. I took the belt off my trousers, my hammock cord and a handkerchief and tied him onto the mule so that he would not fall off again. After a half-hour rest with him on the mule we continued our walk and got out of the forest by night of the fourth day where I collapsed into my hammock in one of the roadside settler's cabins.

That was one time in which sheer will power pulled me through. Yet the hero of this book lived on the brink of disaster with malaria, death of his companions, fighting the animals of the river and finally surviving advanced stages of beri-beri to live and settle down in New York as a tile layer. Just reading these papers is an inspiration as to what one can survive where there is a will. Yungjohann frequently refers to his secret for survival when he had malaria. It was continued action, forcing himself to carry on. He describes his companions who lay down to die. I think that you will find this a story of courage and inspiration.

Ghillean T. Prance

White Gold

*The diary of a rubber cutter
in the Amazon
1906 - 1916*

by

John C. Yungjohann

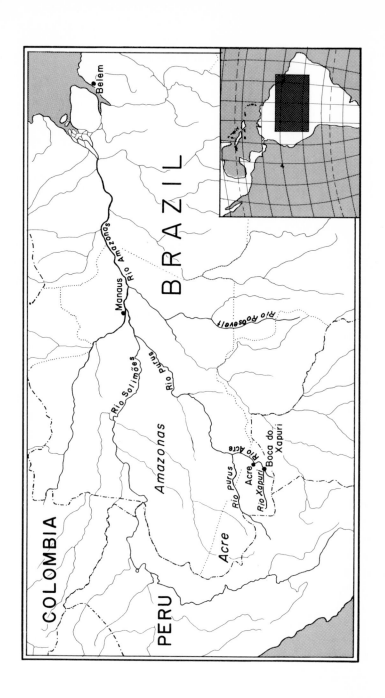

WHITE GOLD

1

The world's consumption of rubber is steadily increasing. In 1906 the total consumed was approximately sixty million pounds of which Brazil had furnished forty million pounds and all signs point to a still greater demand of rubber for it has become an absolute necessity to the manufacturing as well as the medical world. You will find it in every household and very little does the average person know under what circumstances and difficulties the great bulk of this rubber is gathered and brought into the market by risk of life and privation.

As I have spent ten years of my life for this purpose I will endeavor to reveal to the people my experience as a rubber cutter, and also the experience of living ten years in the most dense woods and wilderness of the Amazon Valley, a territory ridden with fever and pestering insects.

Many a mother's son has been lured into that profound wilderness to be kept under bondage, never to return to civilization, only to fall a prey to that treacherous fever beri-beri.

The trip to Pará,[1] South America, from New York takes twelve days, so I landed in Pará which lays at the mouth of the Amazonas river at the end of March.

Getting within twenty miles of this big river the gongs on board the steamer were kept ringing; signs were hung up to advise the people that the ship was sailing through sweet water. The pumps were kept going to clean the boat of sea water and the passengers can have the experience of drinking sweet water in the ocean.

Another feature of this immense river is that it rises and falls every six months. The rising starts on the twenty-first of January until the twenty-first of July, on which day it will start to fall. Steamers going into the interior are leaving Pará with the rising water and will come down the river with the falling water. The reader will clearly see that when I arrived at Pará, at the end of March, I was too late to get passage for the interior.

---

[1] Pará is the modern city of Belém which is at the mouth of the Amazon river and is today the capital of the state of Pará. This story began in March, 1896.

The berries of the coffee plant *(Coffea arabica)*. The beans are inside enclosed by a fleshy pulp.

Coffee beans being dried on the type of flat terraces described.

After landing at Pará and learning these conditions, also finding very little chance of getting employment, I decided to take passage on the next steamer of the Southern States. I got passage on the steamer *Alexandria* of the Uranium Line which was a coastwise vessel to Rio de Janeiro.

Brazil is a Portuguese speaking country and not being acquainted with that language I found it very trying to look for a job. Nevertheless after a few days I got an offer to go to Juiz de Fora a little city of about fifteen thousand inhabitants, located in the coffee center in the State of Minas Geraes. My employer Dr. Antonio Magalhães, was the owner of a good size coffee *Fazenda* (plantation) measuring four square leagues, with fifty acres of coffee trees, two thousand trees to an acre.

Coffee is drunk in every American home but the average people know very little or nothing about the cultivation of the coffee. I worked ten years on the plantation and am in a position to give a good explanation on the treatment of the coffee for marketing.

The city of Belém at the mouth of the Amazon showing the market and the port for fishing vessels.

I mentioned before, the common rule is to plant two thousand trees to an acre, nicely in rows. One man can handle three acres, six thousand trees only. The weeds grow very fast in that hot climate and it takes all of his time to give them three hoeings a year; that is up to the time harvesting begins. The coffee tree does not grow very high, in fact it resembles a bush; a person can easily reach the top branches by bending them a little. The fruit resembles our cherry very much, but is of the more oval shape and does not grow on a stem like the cherry, but grows on both sides tight to the branches. The seed of this fruit is the coffee bean; two of them with the flat sides together enclosed in a very thin shell. When harvest time arrives the fruit gets stripped with both hands from the branches to the ground and runs through a sieve laden in oxen drawn carts and hauled to the *terreiros* which consist of leveled off squares of about one hundred yards, of which they have about four on every plantation. Here the fruit is spread about four inches deep. Two men are kept busy all day turning them with hand plows (if they are not turned constantly they will burn very quickly) until they are perfectly dry, after which they are hauled to the threshing machine to be prepared for the market.

All during the time that I worked there I kept in my mind the fact that I wanted to go for the rubber. In 1906 after ten years on the coffee plantation I returned to Rio de Janeiro, got passage on the steamer *Ceará* of the Lloyd Brasileiro bound for Pará in time to get passage for the interior.

2

The rubber tree is known by the name of *seringueira* and the parts of the woods where the *seringueira* thrives are known as *seringales.* Politicians and men of influence stake as much as they can on these seringales and file a claim with the government.

The state of Amazonas is a tremendous territory; the reader may therefore well understand that these claims take on an enormous proportion; leagues and leagues of those virgin woods are acquired by those individuals by a mere claim that makes them the sole owners. As soon as they have secured themselves that way they put themselves at the disposal of the exporter who in turn will give them an almost unlimited credit. Having gone that far, their first move is to establish an agency in the next best hotel to induce and welcome any young man who is willing to go and risk his life for the money sake.

The Amazonas river is the largest river on the globe with its countless tributaries and as the seringales (rubber woods) extend all over these tributaries there are, or better said, in the city of Pará or the city of Manaus just as many of these agencies. To try and go into the interior without getting in touch with any of these agencies would be next to impossible. In the first place one would not know where to go for the steamer that leaves the city of Pará for the interior that will make its first stop at the city of Manaus, which takes eight days, and from there go directly to its destination. In the second place, it would be just impossible to get a passage, for every inch of space aboard these steamers are taken ahead of time, months and months. They, of course, would not bother with a lonely passenger for they would have to carry his baggage and what's more they would have to carry an additional amount of provision.

Learning of these conditions after my arrival at Pará and being advised that the best seringales were encountered at the river Acre, I got in touch with an agent for that location and put myself at his disposal; which means that he will foot all your expenses and even give you liberal spending money until the steamer sails. I have saved plenty of money on the coffee plantation did not need any spending money; but he insisted on paying all

River launches in Manaus of the type that trade up and down the Amazon River with contemporary rubber cutters.

The floating harbor of Manaus, built in the rubber boom still serves many ocean-going ships today.

the rest, such as hotel and passage. I was also informed not to buy anything in the line of arms, clothes or medicals, for they would buy some wholesale and I would be able to get it much cheaper from them.

When we got to our destination, I found out to my great sorrow why they insisted on paying all expenses and why they advised me not to buy anything. I will relate to this in the next chapter.

After two weeks stay we embarked on the steamer Baturete for the upper Acre region and here is where the important part of my story begins. As I stated before the city of Pará lays right at the mouth of the Amazonas; except for the fresh water, little you see there of a river, it appears more as if the city of Pará was laying right on the ocean. The steamers that navigate the Amazonas are double deckers and resemble the big boats that travel from New York to Albany and Rhode Island.

It is quite a rough trip when these shallow steamers leave the harbor and catch the ocean breezes until you reach Ilha das Onças (Tiger Island), once behind that island everything changes and your eyes behold the most wonderful panorama. The trip across the mouth of the river to Ilha das Onças takes twelve hours and after a three-hour sail we reach the narrows, the navigation channel between the many isles. The beautiful sceneries changes in rapid succession, now and then you see small settlements, yes, even small towns with the church steeple looking out amongst superb vegetation. First we pass the little town of Óbidos and as we lost sight of it we see about a dozen or more wild pigs across our bow making for the other shore. Time and time you will see alligators swimming across. We pass another little town by the name of Santarém, little it differs from Óbidos. From here to Manaus the settlements get scarcer; but, every day you will see new scenes.

After an eight day trip we reach a point where the mighty Amazonas river divides into two branches, to the left the Rio Negro and to the right Rio Solimões. One hours ride up the Rio Negro and to your surprise you see diving out of that enormous entanglement of vegetation an up-to-date city of about fifty thousand people. You see an up to date harbor, ocean liners from all parts of the world, latest appliances for loading and unloading the vessels, the two steeples of the cathedral and the shining copper roof of the opera house. Going ashore you will see up to date trolley cars and also the Hotel International, where our Ex-President

Roosevelt made his stay while he was there after he made his famous trip through that region. Indeed it is a very busy place the biggest part of the rubber from the Amazonas is exported right from here; big wages are paid to the working men and wherever you look the signs show that money is plentiful. But, as everything has an end, the hour of our departure has come.

We embark again and after an hours sail we find ourself engulfed again in that enormous growth of vegetation without any signs of human life. For two days, we amuse ourselves by watching swarms of birds, monkeys, turtles, and alligators moving about, beautiful cranes standing here and there.

We are getting up higher and the supply of coal which was taken on in Manaus is giving out and we are making a stop for taking on wood for fire. Every eight hours we stopped to get wood.

Life aboard the steamer is getting monotonous and as we go up to the river Purus, which is a tributary of the Solimões the water is getting more yellow. The passengers are keeping more to their hammocks and are losing their appetites, you do not see so many at the tables. As we go up the river, turning from the river Purus into the river Acre conditions are getting worse; that terrible interment or swamp fever shows itself and woe to him who has no will power, who keeps laying in the hammock instead of fighting it, his days are soon counted.

After thirty-two days we reach a point where the steamer does not go any further, we had arrived at the mouth of the Xapurí, or as Theodore Roosevelt said the River of Doubt,[1] tributary to the river Acre. We disembark under the impression that we had reached our goal. We find a settlement there of about two thousand people. Anybody is called to help unload the vessel; the captain throws everything on the bank of the river the way they get hold of the boxes, barrels, and packages, there is an absolute lack of system.

---

[1] The actual Rio Theodore Roosevelt; or the River of Doubt, is different from the Rio Xapurí and is a tributary of the Rio Aripuanã which runs into the Rio Madeira. It flows north from Mato Grosso State not Acre. There seems to have been some confusion in his geography by our writer. Both rivers represent extremely isolated remote parts of Amazonia.

It is now up to the agents to straighten their shipments and to get it under cover; everybody is summoned to help, the veterans who happen to be at the Station as well as the newcomers. The strangest thing will unroll itself before your eyes; in order to get the men to carry the provisions away one agent will pay more than the other agent.

The men taking advantage of the occasion will ask for anything that comes to his mind, they would get two and three mil-réis for every item they carry away, which is as much as a dollar, besides all the drink or any other bargain he made. The newcomers do not fare so well as the veterans, as they are not acquainted with the conditions and they are mostly sick and weak with fever.

The Manaus opera house, built at the height of the rubber boom by those who prospered from the exploitation of our hero and many other people.

3

We were two hundred newcomers and after four days of treatment — which consists of swallowing large doses of quinine in powder form, which in turn is put into a cigarette paper and twisted into a round ball with black coffee; also taking large doses of witch hazel and allum to stop the diarrhea which everyone gets from drinking the yellow water from the river — we were told to pick our companions for the season, having made our friendship on the way up, and of course to stick together no matter what difficulties may arise. There were seven of us that wanted to stay together. Next we were told to receive our provisions and outfits, such as arms and the tools for working the rubber.

From the day that I got in touch with the agent in Pará I was treated with the greatest consideration; all I had to do was mention anything I wanted, that is within his limit, and I would get it; but, what came now put everything in the shadows. Putting his arm around our necks and telling us you had better take this and better take that and plenty of it, having no idea of what we really needed, he got us to buy about eight times as much as we really needed, and the price was enormous. I will endeavor to relate some of the prices: *carne seca,* which is salted dried meat is sold at nine mil-réis[1] a kilo, you have to buy it by the *araba* which is sixty kilos that makes 540 mil-réis; araba salt at four mil-réis; sugar six mil-réis a kilo; coffee nine mil-réis a kilo, rice nine mil-réis a kilo; crackers eight mil-réis a kilo; one package of safety matches four mil-réis; one pound can of American lard twenty-four mil-réis; one can of condensed milk twelve mil-réis; one can of sardines fifteen mil-réis; one pound can of Armours corned beef eighteen to twenty mil-réis; a half pound can of Ceylon tea twenty-eight mil-réis; one quart bottle of Port wine eighty mil-réis; one bottle of cognak eighty mil-réis; a pint bottle of Guinness Stout six mil-réis; one kilo of tobacco one hundred and twenty-eight mil-réis; one quart bottle of vinegar eighty-five mil-réis; any kind of pill for fever or anything else one and one-half mil-réis; an ordinary cotton shirt forty

---

[1] Mil-réis was the Brazilian currency when this was written, and there were three mil-réis to the dollar.

Travelling upriver by canoe our author used both paddles and poles.

Struggling to get a canoe upriver on one of the editor's expeditions in 1980 encountering similar difficulties as described in the text.

mil-réis; a pair of working trousers sixty to eighty mil-réis; a Winchester rifle sixteen shot five hundred mil-réis; one box of cartridges fifty in a box one hundred and fifty mil-réis; etc. This is only an idea of how to explain the politeness of the agent and also to describe my feelings when I got the bill for my outfit; it amounted to four thousand seven hundred and eighty mil-réis, I thought I would drop dead.

They next furnished us with a big canoe into which we put all of our belongings. A *mateiro* (guide) was going with us and we were told that we had to go about a couple of miles up the river, so we set off. The places where the river was shallow we were poling, where the water was deep we would have to paddle. We set off in the morning and being told that we would have to go a couple of hours we began to ask if we were getting near our destination, we were always told a little further ahead. We kept on going all day and when darkness set in we demanded to know the truth, how long we would be traveling to get at our destination. The guide had his instructions so he told us that we had not arrived at our destination because we were too slow poling and paddling. There being no moonlight the guide soon picked a place where we could spend the night. The first thing we had to do before it got dark was to get enough fire wood to last all night. This being done we all went in for a bath, each one carrying a large stick, we took turns washing ourself while the others kept on hitting into the water with their sticks, causing as much noise as possible. The reason for doing this is on account of a fish called the *piranha,* they travel in great swarms and are very ferocious. They are of a vermillion color and have very sharp teeth. They are also as quick as lightning and if they got hold of a person their countless number flash by, each one taking a bite. The person's struggles will soon cease, finally to be devoured. Next there is that ever present danger of an alligator lurking about. Having thus taken our bath we prepared our supper which consisted mainly of *farinha* and preserves. Farinha is a coarse flour which serves as bread and potatoes, at the same time. Having had our supper we fastened our canoes with a long vine, which we had procured when we looked for the fire wood, let it slide well into the river bed to keep the canoe in the current.

It is a common belief amongst people that the alligator will attack anywhere but it is not true, he will never attack where there is any current;

in fact, he will never venture into any current except when he crosses from one side to the other side of the river. Thus prepared we passed our first night in the open; each one laying down where he could find a place.

The days are very long and the nights are short under the equator. At four o'clock in the morning it is daylight. After having our breakfast we started poling and paddling again; when tired and disgusted we would throw our pole or paddle down and ask how much longer we would have to go; we always got the same short answer "a little further ahead"; thus we went on day after day.

When we got well up the river we had to face difficulties of various kinds. In the first place we were all sick with the fever and this together with the strain of pushing the heavy laden canoe against the current, no appetite for we were all weak and disgusted; sometimes we were all laying down with the terrible chills which they call *seceoa* in Portuguese. This fever starts in with violent chills to such an extent that every joint of the body pains from shaking, which last for an hour and a half, after which violent headaches and heat sets in which last for about twelve hours. If a person has not got enough will power to fight it, utter collapse will follow. In some cases it will appear every day for three to four days, very rarely in five days in which case it means death. In other cases it appears every other day three or four times and then again a person will get it once or twice a day for perhaps five to six weeks. As mentioned in the foregoing

Caiman basking in the sun is an all too rare sight in today's Amazon, but was common when this account was written.

chapter, large doses of quinine is the only thing to keep it down. Very often the patient will lose his reason and will run off into the jungle, unconscious of what he is doing; and very peculiar they can never be found. I have never heard of a single case in ten years of my stay there where one had ever been found.

When we were on the river mosquitoes did not bother us much but, instead, there was a much worse pest, the *pium,* a tiny little fly, gray with yellow spots, which would appear from daybreak until the sun was very hot, say about nine o'clock, and evenings from about five o'clock until dark. It gives a terrible sting and not even tobacco smoke will drive them away.

Sometimes the river would get very shallow at which time the canoe would strike bottom; sick or not sick we would have to jump into the water and push the canoe ahead until we got into deeper water and so on. During the day we would have a torching heat and nights were chilly to such an extent that although wrapped up into two blankets we were shivering with cold. The scenery was simply stupendous, something that can never be forgotten. Enormous big trees, palms, bamboo and bushes, every thing in chaos, growing to the very edge of the river. The ever present smell of flowers and blossoms would at times get so strong that it would be sickening. Early in the morning the trees were alive with parrots, screeching and whistling, so that we were compelled to yell to make each other understand. Swarms of various kinds of monkeys would come to the very edge of the river gibbling and then disappear. There were turtles on both sides of the river and now and then we would take a shot at an alligator. Early in the morning we would scare up a herd of *capivary*[2] (river hog). They would jump right over our canoe into the water making a terrible splash; first this gave us a mighty scare but soon we got used to it.

The overhanging trees were alive with birds about the size of a Guinea hen called *sigano;*[3] flying up and screeching, causing a terrible comotion

---

[2] Capybara *(Hydrochoerus)* is the world's largest rodent which can weigh up to 100 pounds and is a frequent riverside animal in Amazonia.

[3] The bird commonly known as Cigana in Amazonia is *Ophisthocomus hoazin,* the hoatzin, but the horned screamer *(Anhima cornuta)* fits the above description much better and is known locally as *inhuma.*

A capybara, the world's largest rodent..

The house of a rubber gatherer.

as we would approach. They lay a good size egg, hundreds and hundreds in one nest. We would get our daily supply of fresh eggs from them; the birds are not good to eat and to get their eggs was by no means an easy task for wherever there was a nest it was built on branches hanging over the river and directly under the nest there would be a wasp nest of about three feet in diameter. The wasps themselves were about an inch long, black and had a terrible sting, the least touch would start them buzzing. The only thing to do then was to jump into the water and dive in whatever water there was, even then they would get us so that there was always someone with a swollen arm or face.

The Amazonas with its tributaries is alive with fishes[4] of so many kinds that I do not think there is a man living that has seen them all. I will endeavor to name some of them. There is a bull fish, a full size one weighs about five hundred pounds, it got its name from having a head like a bull; he is a very tasty fish and is much sought for. The *Piarague,* or sweet water Sturgeon, weighs as much as six hundred pounds. The only way to catch him and the bull fish is by harpooning. The meat is cut in slices, salted and dried in the sun to be roasted when wanted. There is the very tasty *Bagaxe,* a fish of about twelve to fifteen pounds; he is caught in great quantities by the Indians as well as by the natives. There is the *Pirapitinga,* equally as good as the Piranha which I have mentioned before.

After forty-five days counted from the day we started from the Boca (mouth) do Xapurí, we were told by the guide that we had reached our destination. Although we had gotten used to the traveling in the canoe,

---

[4] The writer refers here to five types of fish. The bull fish is a translation of the Portuguese, *peixe-boi,* which is the manatee. This is actually a fish-like mammal *(Trichechus inunguis)* that used to be abundant in Amazon waters, but is now on the verge of extinction. The pirague, judging by the description, must refer to the pirarucu *(Arapaima gigas),* the giant lung fish which surfaces to breath and thereby gives itself away to the waiting harpoonist. It is still an important source of meat in Amazonia. We are unable to identify the name *bagaxe.* The pirapitinga is the characoid fish *Colossoma bidens,* closely related to the tambaquí in the same genus, and is the second largest scaled fish in Amazonia. This fruit eating fish is one of the best tasting of all Amazonian fish. The piranhas are the well known flesh eating fish whose danger has been excessively exaggerated.

we were nevertheless glad that we had come to a stop. It was thirty-two days from Pará to the mouth of the river Xapurí, therefore, it made seventy-seven days that we were traveling.

A pirarucu fish and the fisherman.

4

Having arrived at our destination we immediately set to work to get our cargo in safety. With no sign of human life about us the reader may understand that we stood there in the midst of an almost wilderness. The first thing was to look for and clear a place where we would build a house big enough to hold the seven of us and our cargo. This was no small task considering the chaos of monstrous trees, thorny palms, bamboo, and brush wood. Having succeeded with this under the instructions of the guide, we then had to build the house, which was quite simple as far as the structure was concerned. However, the putting on of a waterproof roof was a little more tedious. We had to gather a certain kind of palm leaf, then braid them on bamboos of about ten feet long, these in turn were put in layers on the roof, similar to our shingles. Next came the flooring which was put about four feet above the ground. To get the flooring we had to cut down a certain palm called *bacaba,*[1] which grows very tall; cut them in desired length, split them on one side and after turning the split down we would hit it with the back of the axe until the whole length would split and crack so that it would open up flat. The white corky stuff that forms the inside of every palm gets cut out.

Having completed this task we then went to work and carried our provisions and outfit to safety. Up to the minute that we had finished this, the guide had been very friendly, always willing to help and advise us in our work, as well as when we fell ill with the fever, but from now on he showed an all together different attitude; he became very sullen and would only speak when asked for advice. We soon found out why, for no sooner had we gotten our provision in safety, he set the canoe adrift, which meant that we found ourselves out of civilization entirely. We realized our position but doing all this work besides suffering with the fever we had very little will power and let things go as they came.

From the day that we had arrived until now was two weeks; allowing us one more week for rest, the guide took us out to instruct us to our task

---

[1] Bacaba is a palm of the genus *Oenocarpus,* the wood of which can be easily split into planks for use as a durable flooring to houses.

of gathering rubber. It was high time for us to get a start for the season only lasts from the middle of April until the middle of October. We had about two weeks work to do before we actually could tap the rubber tree. First we were instructed how to recognize the rubber tree which was entirely different than the ones they raise here in the flower pots. There are two kinds, one has a smooth white bark similar to the birch tree, the other has a rough pink bark; both grow very tall and have the same leaves, each leaf having the shape of three fingers. As it is impossible to locate the tree on the ground on account of the heavy undergrowth, they can only be located by their leaves.

Each one is armed with a rifle and *facão,* which is a heavy broad knife about twenty-four inches long, (made by the Collins Company, Collins-ville, Connecticut). We set out from our house, which by the way is called a *baraca,* and we cut a small path from rubber tree to rubber tree zigzagging on as you spy a tree, but in such a way as to form a loop or an eight, returning always to the point where you started, your house.

As a rule a man works from eighty-five to one hundred and twenty trees, according to the size of the tree, for in some places one will encounter all heavy trees while in other places it will be just the opposite. Then again it depends a lot on how far they are apart which varies much in different places. The paths are called *estradas,* each man works two estradas, in one today, the other tomorrow and so on. We being seven men made fourteen estradas that we had to cut, through heavy bamboo groves.[2]

The bamboo grows to the very top of the highest trees, varying from five to two inches. You cannot cut between, they must be cut off at the ground, lifted up and pushed aside. They are full of sap or water, which by the way is always cool and very good for drinking.

It is by no means an easy job to get a path cleared through these bamboo groves. Then again the scenery changes, we encounter *mumurú* groves. The mumurú is a palm, trunk and leaves are completely covered with thorns, on the trunk four inches and on the leaves between one and two inches long. They give an abundance of fruit, they resemble very much

---

[2] The presence of such tall bamboos in the forest is characteristic of the forest of Acre and nearby Peru, but does not occur elsewhere in Amazonia. This region of bamboo forest is dominated by bamboos of the genus *Guadua.*

our peaches, they are also covered with thorns so that it becomes a task handling them. They do not grow very high, very rarely above fifteen feet and about nine inches in diameter; but, they grow very close together.

We had to go through swamps and creeks, where we have to cut trees down to serve as bridges. I would like to mention here that we pass over the same creek time and time again. Many times nature provided bridges in the form of vines that grow from one tree to the other. A monstrous growth of vegetation is striking in its greatness and closeness. We are not given much time for thoughts as the guide urges us on in our work. The terrible fever has shaken us up and we are too weak and tired to care much about scenery.

Having finally opened up fourteen estradas the guide gives us the final lesson, how to tap the trees. Then the most particular part how to smoke the milk to form the *goma elastica.* In our outfit each man got a crate of tin cups; each crate holding one thousand cups. These cups are of a cone shape, measuring four inches high, three and one-half inches at the top and two inches at the bottom. These cups each man distributes along the estradas that he has to work, leaving at each rubber tree the amount he needs; if the tree measures four feet in circumference he leaves four cups, if it measures five feet he leaves five cups, and so on.

Each man also got a little steel axe the cut of the axe being about two inches wide; on this axe we put a handle about ten feet long and we start to bleed the trees. The first six days (three days on each estrada) only water runs from the tree cutting them every day; after that the milk comes, resembling in color and thickness our evaporated milk.

As I mentioned before we put a handle on the axe of about ten feet and reach up and give the tree a slant cut, then take a cup, which is called *tijehle,* jam it into the bark as high up as a man can reach directly under the cut, and the milk will run right down into the cup. This is done a foot apart, around the tree. You go along the estrada until you come back to the starting point. You start at three in the morning until about six at which time a bath is taken and something to eat. You are then ready for the second part the gathering of the milk.

Each man gets a tin bucket in his outfit, narrow at the top and wide at the bottom, resembling our milk cans, which hold about fifty pounds, twenty-five kilos for they count their weight in kilos. The guide shows us

The trunk of a large, old rubber tree.

Rubber latex being collected in a bowl.

A partially prepared ball of coagulated rubber.

how to make a crip for carrying on our backs, into which the bucket is put, and off we go on the second lap, from tree to tree. You take the crip off of your back and empty the cups in the bucket. Some of the cups are half full, some three-quarter full, others running over. In the last case you would put an additional cup below to catch the overflow. The gathering in of the milk takes considerable time. A man has to be pretty lively to be back at the starting point at six o'clock in the evening, carrying from forty to fifty pounds beside the rifle, edging his way through a tunnel-like path hardly two feet wide.

Having gotten back safely there is awaiting two more hours of hard work, the process of smoking the milk from the goma elastica, which is done as follows: a coop has been built, one for each man, about the same way we would build a chicken coop, two slant sides heavily covered with palm leaves, taking in a space of ten feet by ten feet. In his outfit each man gets a conical cylinder, having an opening at the bottom of about three feet and at the top three inches. A hole of about eight inches is made in the center of a coop over which the cylinder is put. Two cross-bars are

Smoking of rubber to coagulate it into a large ball.

The author describes his rubber smoking shelter like this one as a "coop".

A calabash on the tree before it has been dried for use as a utensil.

made, one on each side of the cylinder, about a foot higher than the opening of the cylinder. Along side of this arrangement three poles are hammered into the ground in such a way to carry a round basin big enough to hold the quantity of milk gathered. The coop is small and the man who does the smoking has to sit right in it. He has to have wood that produces smoke that will not choke him or burn out his eyes; nature provided that also.

The guide calls all hands together and armed with rifle and axe we dive once more into the growth of freaks and curiosities. Soon the guide finds what he wants, a big tree called *garape*.[3] This wood is orange-yellow and very hard. As night is drawing near all hands start to chop away at the tree for dear life for none of us was handy with the axe. In about three-quarters of an hour after a lot of sweating and fretting, we succeeded in getting the tree to tumble over with a crash. After cutting off enough shavings for every man we make back to our coops.

Fire is made in the cylinder and the green shavings jammed in tight with a stick. A few minutes later and a heavy white smoke comes pouring out, then a stick about two inches thick and five feet long is put across the two bars directly over the opening. With a small *calabash* [4] shell the milk is slowly poured on the stick with the one hand till all the day's harvest is smoked in a ball around the stick, taking great care that no flame comes out for that will set the rubber on fire immediately and a whole days work would be lost. As a man goes through the same routine every day he very soon becomes an expert at his work.

---

[3] I do not know of a wood called garape in Amazonia. The nearest to this is *caripe* which refers to various species of *Licania* which are sometimes used to produce smoke for rubber coagulation and is a very hard wood. The smoke is often produced from the wood of either the massarunduba *(Manilkara huberi)* or acapú *(Vouacapoua americana)*, but most often used are the oily fruits of various palms, especially *Scheelea martiana.*

[4] The calabash is the most commonly used gourd-like utensil used in Amazonia and it comes from the tree *Crescentia cujete* which produces the soccer-ball sized calabashes on the trunk and branches. The inner pulp is scraped out and the outer shell is used.

WHITE GOLD

5

Having described the work of a seringueiro (rubber cutter), I will now relate some of my experiences. I mentioned in the foregoing chapter that there was a change in the behavior of our guide after we had our provisions safely ashore. After we had our first days work done the change was still greater. No sooner did we have a good fire made for the night he disappeared into the dark without taking the usual bath and without eating his supper and never saying a word. We in turn took to our hammocks. Early the next morning finding that he had not returned we went to our task as good as we could, with the result that four of our companions got lost right on the first trip while cutting trees. In such a case we all kept signaling with our guns which is quite an expensive thing for every shot cost three mil-réis, besides losing the day's work. This by the way, is a daily occurence, that one or more of the company gets lost until we get accustomed to the woods.

The evening on the third day the guide unexpectedly returned, carrying a big pig on his shoulder. Throwing it on the ground he told us to take the skin off and roast a piece for supper. While we were busy doing this he disappeared never to return again. We settled down to the fact, however, and went along trying to make the best of things; but, with very little results, for either one or more would get lost on his trip which, although it sounds funny, happens very easy. It sometimes happens where the estrada makes a sharp turn and a man has to stop and tap a tree; after he is done he starts to walk in the opposite direction, everything looks alike and he gets bewildered. Very often a thunder storm passes by and the heavy rain closes in the estrada for perhaps fifty feet or so. A man comes along and trys to get through, if by chance he hits the estrada right again he is all right, if not he will not find his way again unless others come to his help. Sometimes it happens that he steps off his estrada to shoot a prey which he spies, or the fever throws him down, here is where the will power counts. I might add here that it took me fully two years until I was able to go into the jungle for days or weeks and then find my way back home.

Sometimes all hands would be laying in the hammocks shaken with the fever; one would be calling for water, the other for hot tea and the other for blankets and so on. I myself being just as bad would get up and stagger around helping the others as good as I could. They soon got used to that and demanded it from me. Nobody wanted to get any fire wood for cooking or for the night fire and no one wanted to go for drinking water; everything was left to me, which soon proved to be my salvage.

After being in camp for about eight weeks one of our companions died, a young man by the name of Andrew Benton; all I know of him is that he came from Cincinnati. This made a terrible impression on all of us. None of the others could give me any help to bury him and I myself was too weak to carry him out, so I set to work to dig a hole in the floor just below his hammock. Having no pick or shovel I had to do this with my machete and hands. I cut the countless roots and loosened the ground with the machete and threw it out with my hands, a hard piece of work for a sick man but I succeeded at any rate, to get down about three feet. I then cut the hammock so that it fell with the body right into the hole, I closed it up and replaced the flooring.

I do not know what happened for several days, but, when I came to I was confronted with another job, but this time it was two that I had to bury; how I ever did it I do not recall. I got them covered in the ground somehow. After that things got a little better and we all rallied a bit in such a way that all four of us were able to work again.

The first thing we did was to move our house. After that we went at the rubber again; but it did not last long before we were all laid up with the fever again. One right after the other my last three companions died. Being too weak to do anything I simply cut the hammocks and let their bodies drop to the floor. Leaving them on the floor, I dragged myself to one of the smoking coops and made my stay there. After several weeks (days and months I had lost track of) I was able to go to work again. Taking a peep over at the house one day, where the three bodies were laying, to my surprise I saw the skeletons laying there as clean as if they had been going through some cleaning process. I carried them some distance away into the bush and went back to live in the house. I got good and strong and was able to go to work every day, but now something else presented itself. The Indians started to bother me.

As long as we were all there we had never seen one, although the guide had warned us, but as long as we never saw or heard anything of them we forgot all about them. Coming back from my trips I noticed several times

Yanomami Indian approaching through the forest.

that somebody had been at my house. Although there were no footprints I first thought that the guide was crawling around, but, one day I caught a glimpse of two Indians as they swept by. That settled the case, I knew what I was up against and of course I felt very uneasy. I kept my eyes opened from then on, but so did the Indians. Seeing that I was alone they showed themselves more and I got a chance to see what they looked like. Coming home one night I found that they had set my house on fire, but as everything was damp it had not done much damage. I had to stay home the next day to fix things up again. A couple of days later they did the same thing, this time with better results. The smoke spoiled all my store of farinha, rice, sugar and coffee, the canned goods of course were not harmed. Besides they had taken away all of my rubber and the little bit that my companions had made; that was a hard blow. I was deadly afraid at night for fear they would do me harm while I was asleep. This was

Mayongong Indian community dwelling house.

ignorance on my part as they do not travel at nights as I found out later. It was no use for me to make rubber if they would steal it on me, so I made up my mind to get in touch with them and find out what they wanted. I walked along the estrada and when I spied any of them I would put my rifle aside and motion to them to come. After doing this several times, one day they motioned for me to step up nearer to them. I did this and some of them stepped up from behind me and grabbed the rifle and closed in on me from all sides, there being fifteen of them. They then walked me into the woods.

This happened around nine or ten in the morning. They marched me all day till sunset. They stopped for the night and the campfire was made. Some fowl that they had shot was quickly cleaned and roasted. I was given a good share. After eating we stretched around the fire for the rest of the night and I was no further molested. They knew very well that I could not run away. Early at daybreak the next morning they started off again without anything to eat until well in the afternoon, going at a fast clip all the time, edging, stopping and jumping continuously. When we reached their camp which they called *maloca,* a big swarm of men, women and children came out gaping and talking. It took all the attention of the men that brought me in to keep them away from me. They did not seem to be having any bad feelings towards me, at least their actions did not point that way. One of the men that brought me in disappeared into the tent which I did not see anything to report. I later found out this tent belonged to the chief which they call the *Xoa-Xoa.* Having gotten the order to bring me before him I was led into the camp amidst the swarm of people, but when we got near the chief they fell back and formed a ring around us and were as quiet as a mouse. The chief was a short man of about forty and to my great surprise he asked in a very broken Portuguese what I wanted. I told him his men had put fire to my house and stole my rubber and that I would like to have it back. Also for him to tell his men to leave me alone. In answer he snapped back, *"cala boca",* which means "shut your mouth." He motioned to sit down and I was given a good supper, shown a place to sleep and that ended the whole thing for that night.

6

The Indians, of which I am speaking, belonged to the Tupí tribes,[1] and I will describe to the reader their life and habits. I mentioned before that they call their camp a maloca, in the Tupí language. It consists of a big open barn posts and roof, about three hundred by forty feet, very nicely done, the whole floor is covered with white clay which they call *tabatinga,* which is well kneaded spread over the place and polished with the bare feet until they get a shine to it. This big barn is divided into three places, the first place serves as sleeping quarters, endless rows of plugs sticking about six inches out of the floor and gives the place a rather uneasy feature. By closer examination one finds them all systematically arranged, in groups of eight, four feet apart and nine feet in length. Over each of these groups hangs a bundle of mats and rods. When the time comes to retire you take the rods and tie the end with two little rings, which are made of what they call *baust,* which is taken from under the bark of a certain tree and it is very tough. This done they bend the rod across to the opposite plug fastening the rod there in the same manner, thus forming a hoople. This is done with the rest also until they have four hooples made, they do this very swift. Over these hooples they spread the mats also putting some under the hooples on the floor. This now looks like a tunnel, four feet wide by nine feet long, into which they crawl for sleeping.

I cannot well describe the impression I got when I was brought into camp. The men and women do not wear any covering whatsoever. I was afraid to look around in fear some of the men would jump at me and cut my throat. I kept my eyes glued to the chief until I was shown a place to sleep and wondered what the next day would bring. But in the days that followed I learned different. When I awoke at daybreak on the next morning I found the whole maloca deserted. I learned later that they do this every morning from about half-past four until seven o'clock. When the sun gets warm they come out and dry themselves in the sun. They then

---

[1] According to Steward's *Handbook of South American Indians,* the Indians that inhabited the upper Rio Xapurí were Ipurin and Ainamori, both Arawak tribes. Tupí is one of the major language groups of Indians and the nearby Catubina Indians have sometimes been classified as Tupí speaking. It is unclear what tribe our writer visited.

The author describes his uneasiness at first among naked Indians.
These two Yanomami are painted for a feast.

The roots of the cassava or manioc plant.

all come in for feeding. They are a clean healthy people. In their sleeping quarters the married couples with their children sleep in the center, the single men on one side and the single girls on the other side. When they arise in the morning the first thing they do is to tie up their mats and rods and hang them on their place. They then go for their bath.

Feeding time with them is the most interesting thing I ever saw. The men go hunting and fishing and give their catch to their wives, mothers and sisters, who in turn prepare it for eating; all is made the same way. They stick a thin rod through the meat or fish, whatever it is, and turn it over the fire until it is roasted. The children are fed first. They sit down on the floor cross-legged forming a big ring, as quiet as a mouse. The mothers then go inside the ring and tear the meat or fish apart with their hands giving each child a piece until they are all fed. The children then go out and the grown people sit down and eat the rest. However, they are not such great meat eaters as one would think they are, they eat a lot of fruit in between. They have big banana groves and they also grow an enormous lot of mandioca root, the roots are gathered and mashed up in big troughs, the mash is put in baskets so that they can press the juice out of it, then it is spread over clay ovens about two feet in diameter and about one-quarter of an inch thick, and baked. They call them *beijú*. They make them day after day and eat them any or all times during the days. They

One of the cassava products much described in the text is this flat cake called beijú.

The cassava roots are poisonous and after soaking all night this apparatus is used to squeeze out the poisonous juices.

The grinding of the cassava roots on a grater made of a board with embedded stones.

The final part of the preparation of cassava flour is roasting on an open oven with a fire underneath. It is stirred to prevent burning.

also plant a lot of corn which is entirely different from our corn here, it is dark blue and each ear is about two feet long, but it tastes the same as our corn. When the harvest is near ripe, they mash the corn up in troughs mixed with water and put into big urns which they make themselves and put to ferment.

They have rather a peculiar way to start the stuff fermenting; they take four or five of the favorite girls, each girl gets a half ear of corn, they then sit down on the floor putting a small pot between them, and start to chew the corn off and spit it into the pot, when they have the desired amount they stir it into a mash and let it ferment. After two days it is ready to drink with wild honey to sweeten it. They call this drink *mixire.*

Fruits grow plentiful in the woods, I will name some of the more important ones.[2] The *monga,* an oval shaped fruit about the size of a big apple and very delicious in taste; the *piquiva* looks exactly like a yellow plum; the *yacadicaba,* the most delicate fruit of all, they look like a yellow plum and grown on the trunk of a tree and on the roots that are exposed from the ground. There is the *murumurú* which I already have mentioned in one of the foregoing chapters; the *cacao* fruit, the seeds of which gives us the much liked chocolate; there is the *castanha,* a long v-shaped nut, which is well known as the Brazil nut. Then there are two very important palm fruits, not so much to their taste but for their highly nutritious quality — the *açaí* and the *bacaba,* each palm bearing bunches so big that one man very seldom is able to carry one bunch. When they are ripe they are black and as hard as a stone, they are put in hot water for about twenty minutes to soften up, then they are mashed and run through a sieve. The juice thus gained has the color of oxblood and tastes like butter milk. Each night they pass their time at athletic games, such as running, jumping, wrestling and climbing, the Xoa-Xoa giving the decision. They are very hospitable, except on certain days, when they hold their feasts. On these

---

[2] See *Notes* at end of this chapter.

The fruit or pod of the cocoa or cacao tree *(Theobroma cacao)*.

Brazil nut fruit on the tree. The nuts are arranged inside these round fruits like segments of an orange.

days they become dangerous. There were about six hundred grown people in the maloca and for these they prepare great quantities of mixire and beiju (the mixire is intoxicating). Next they prepare a lot of snuff[3] which they call *po,* this snuff is a purple powder ground fine, between flat stones. They put this in tiny baskets, coated with wax so powder will not fall through. Between their dances they sit down and drink mixire and putting small bamboo canes into this snuff, blow it up each other's noses. They keep this up for hours and they seem to lose their reason. With the flames of the fires shining on them, painted up as they are, they look and behave like devils, men and women alike. Before they start all their bows and arrows are piled up alongside of Xoa-Xoa, who does not take part, but sits there like a rock giving signals to start and stop the dances. No matter how wild they would get, they were taking the greatest care not to get closer than a certain distance to him.

When a young Indian has chosen his mate he has to ask permission from the Xoa-Xoa who will then hold counsel. Having agreed, the Xoa-Xoa will hand the young Indian a bow and a certain amount of arrows which he himself made and is sent off into the jungle not allowed to return unless he shoots a puma and a pig, a sign that he is an able hunter and supporter. Should he return without this prey he is for all times marked to stay single and shunned by them all. They are great marksmen with the bow and arrow, a great stunt is to lay on their back and pick a bird from the highest tree.

When they go fishing they cut a certain vine which they call *cipó;*[4] these they club with a heavy stick until they are good and soft, after which they

---

[3] Many Indian tribes have hallucinogenic snuffs made from various plants. The author does not give enough details to know the botanical source of this snuff. The commonest purple colored snuff comes from the bark resin of the *Virola* tree which contains highly active indole compounds related to LSD, and is still used widely today by many Indian groups.

[4] Cipó is the Portuguese for vine. The vine which is most commonly used for fish poisoning by this method is *Lonchocarpus utilis,* a member of the Legume Family. The Amazon Indians still tie bundles together in the stream exactly as described here. The fish are asphyxiated because the poison interferes with their gills rather than becoming assimilated in the body, hence there is no danger to the person eating the fish. The commercial insecticide rotenone is extracted from the *Lonchocarpus* vine.

Indian child practicing with a blow gun. They learn to use their weapons with great skill at a young age.

Mixire offered by the Mayongong Indians. The fermentation is started with human saliva.

are tied in small bundles then they form a line across the stream with their canoes; two men in each canoe one keeping the canoe against the current, while the other keeps batting the water with his bundle. The water gets all soapy from this about a half a mile down stream. They have formed another line across the stream the fishes that are in between, soon come floating along and are picked up. Although the fishes are stunned from this vine they leave no bad effect on people. I myself have eaten them many a time while I was visiting them.

Notes

Eight different fruits are mentioned, most of which can be identified:

Monga is probably a mistake for *manga,* the Portuguese for mango. Since the mango is a native of tropical Asia introduced in post-Columbian times, there is some doubt that it could have already been in such a remote part of Acre; however, the description fits the mango well.

Piquiva is probably the piquia or *Caryocar,* a fruit which is widely used in Amazonia both for the outer pulp and the oil contained in the seeds.

Yacadicabais the Myrciaria cauliflora which does indeed bear its fruit on the trunk and main branches. The delicious edible fruit tastes rather like a grape

Murumurú is the spinous trunked palm Astrocaryum murumuru.

The *cacao* is the chocolate plant, *Theobroma cacao.* The cacao beans are surrounded by a fleshy pulp which is a delicious fruit.

Brazil nut, *Bertholletia excelsa,* is a huge tree which produces large round fruits larger than a baseball with the nuts arranged inside like segments of an orange.

Açaí is the *Euterpe* palm whose purple colored fruit are much used as a drink and ice-cream flavor in the state of Pará, Brazil.

Bacaba is the *Oenocarpus* palm whose fruit is used to make a drink. Like the açaí it is most nutritious and therefore important for its food value.

7

Having given you a fair illustration of the Tupí Indians, I will continue to relate my experience. After being about two weeks with the Indians the Xoa-Xoa asked me if I wanted to stay with them or go back to my house. I told him I would like to go back and I would also like his men to give me my rubber. He called a young Indian who seemed to be very intelligent, and whom the Indians called Xiko; the Xoa-Xoa told him to call some men and bring me home. In a very short time we were on the way to my house, they gave me back my rifle and made me walk in front of them, which made me suspect that they might try to do something to me, but nothing happened. When we reached one of my estradas they left me and I got safely back to my home. After that the Indians often stopped in on their trips and had some coffee which they like very much.

The fever did not bother me so much now and I went to work the best I could. My rubber having been returned to me by the Indians, my main trouble now was the Boa constrictors, they were visiting me pretty near every night. Even the fire did not keep them away and I was afraid to go to sleep. They are very big, the fat part of their body is about seven inches or more and between fifteen and eighteen feet long. They come crawling along noiselessly and roll their body up alongside of the hammock laying their heads on one's chest, from time to time sticking their tongue, which is three to four inches long, in the corner of one's eyes and mouth, then again sticking their head, which is ice cold, under one's arm pit. If you would make a move they would blow like a cat, in fact they do that every time they hear a noise on the outside.

I am wondering what was to be the end of this all but everything comes to an end. The season ends as soon as the rain sets in, which is about the middle of October; so one day coming home from hunting, to my surprise and joy, the guide had come back and had already saw the situation, that there was only one left of the seven he brought there. Which of the seven was left was immaterial to him. Taking it easy for a couple of days, we started to carry the rubber to the river and tied it together which is done as follows: when the rubber is smoked in a round ball over a stick about two inches thick, a size when it gets too heavy to handle, say about sixty kilos, the stick is pulled out and the ball is rolled in the open to dry, thus leaving a hole through the rubber ball. Through these holes the rubber is tied together with strong vines to make sort of a float, the guide giving me instructions.

Leaving everything else behind except my rifle I set off floating down the river, using a bamboo pole to keep the float in the middle of the river. What I had been going through was nothing compared to what was coming now.

First of all, my float was very shaky; it was impossible to tie the rubber balls so that they would not roll a little. I knowing what was in the water had to do stunts to keep balanced, and still I took tumble after tumble in the water. It soon dawned on me that it was impossible to keep that up day and night so I picked out a place to stay over night; I fastened the rubber and made a big fire. Of course, I would not dare to go to sleep during the night and the next morning I went hunting for my breakfast and then started off again.

The start had been very good as the river was very low. The first thing I knew I was caught in one of the big trees that had tumbled into the river and the rubber disappeared from under my feet. I made for the shore

A rainbow boa constrictor, one of the Amazonian boas. We do not know which sort visited our author's hammock at night!

WHITE GOLD

holding my rifle in my one hand. When I got on shore I saw my rubber coming to the surface one by one. Now I had to tear along the river edge like a wild man, through thorns, mud and bushes, until the next bend of the river. I plunged into the river and headed off the rubber bringing it ashore one by one, going down the river all the time while doing this. When I got them safely ashore, I had to back up along the river where I landed the first one and where I had left my rifle. I tied them together one by one as I went down towards the last one and was on my way again. Now and then hunting or fishing for something to eat or to get some fruits, only to have the same thing happen over again; only with the difference that some of the rubber got caught under the water I would then have to dive time and time again to locate it and get it loose. To get some sleep I had to pick my places where I could tie my rubber so that it would stay in the current.

When I left it was near the end of October and it was about the middle of March when I got down at the mouth of Xapurí, the stopping point of the steamers. The trip taking me four and one-half months. Here the rubber gets weighed and the steamers bring the New York and London prices. So much you owe me and so much your rubber is worth; if you do not make enough rubber to pay the bill you are in debt, so it was my case.

As I mentioned before, the season to harvest the rubber is from the middle of April until the middle of October, I was therefore a whole month late and had no chance to make friendship and pick companions for the next season. Being in debt and late I was urged to pick amongst those that were in camp and start right back again. I being in fair health made up my mind to lose no chances and time to get enough rubber to pay my debts and get back to civilization. With two companions I started off paddling our laden canoe up stream, this time without a guide. My two companions were two big strapping fellows, one an American coming from the west, and when I asked him what place he came from and what was his name, he just said, "I am a western boy and just call me Joe". The other fellow was a Portuguese, a very braggy, noisy fellow. Whenever the fever would let up on him he would cave in by the slightest attack, as big as he was, although he had the experience of one season behind him the same as myself and Joe, most of the time he was of no use to us, and we knew we had a burden on our hands. The outcome was that when we got up the river about thirty-five days we would not carry him any longer and we could not desert him in the wilderness neither. We made a raft and let him float down the river again, leaving him to his fate; we heard later that he was never seen again.

Joe and myself continued on our journey finally getting back to the place where I had my first experience as a serigueiro (rubber cutter). We did not lose much time preparing for there were fourteen estradas that we could pick out to work on. Being both in fair health, we worked day and night up to the end of July.

I would like to make a little explanation about night work before I go on any further. Of course, we could not work more that two or three days a week for that would kill the tree working nights. We would start about six o'clock in the evening on the first lap, carrying a torch of rubber after dark with us. The torch is made the following way: a green stick is split up on one end then a piece of rubber big enough to last the trip is wedged in and tied up, this serves as a light as well as protection. The trees will bleed in the evening and will run all night. The dew will help to keep the cut open whereas in the day the sun will dry the cut closed. Each tree will yield at night three times as much and will kill the tree if overdone.

Towards the end of July, my companion Joe got a violent attack of fever, which after two weeks ended in beri-beri *(galapanta)* hasty, which only lasted one hour and again I was left all to myself. Whenever I was lonesome I went to stay a week or so with the Indians, doing everything to make good friends with them. The Xoa-Xoa gave me the privilege to pay them a visit whenever I wanted to. On these occasions I would teach the children different plays that I remembered from boyhood. If there was any amount of beeswax, I would teach them how to make candles using bamboo as dies for casting. Since they have nothing else to light up their maloca than the fire, they were very quick grasping the idea. After this whenever I would visit them, they would have candles all over the place at evenings. On these visits I noticed that whenever they had an attack of fever they would work in the sun until the chill would ease up. I followed this same method and soon found out that by doing this I would get over with much more ease than if I would lay down in the hammock. The results were that by doing this I could attend to my work better and when time came to start down the river I had a nice lot of rubber together with that from my companion Joe. The trip down the river was about the same as in the first time, I had the experience of the first trip so that it took me only half the time. I arrived at the Boca do Xapurí about the middle of January. When I arrived at the station I had my rubber weighed, and I learned in despair that I was still in debt.

I went over to the camp to mingle with the boys and I soon found out that with few exceptions they were all in the same boat. Everybody was gambling and drinking, in fact I beheld a sight when I entered camp which I will never forget. They all looked like a lot of cut throats, long beards, long hair, rifle in hand, facão hanging on the side and a pointed dagger sticking out of the *lumbrigado,* a long six inch wide bandage, which is wrapped around the waist, all bare back, wearing only a pair of cotton pants, a pair of self made rubber shoes and a cap to keep the hair from catching in the bush. The funniest thing was that I looked exactly the same as the rest. The agents give the men all they want to drink and money to gamble.

Balls of crude rubber ready for trading. These with the brand mark of their owner are the final product of a rubber cutter.

8

There are two kinds of rubber exported from the Amazonas, the *boracha* (goma elastica) gathered from the seringa tree and the *kautxuck,* gathered from the Kautxuck tree.[1] The Seringa or rubber tree can only be worked in the dry season, where the Kautxuck tree can be worked all year around, but with the best results in the wet season. The Peruvians work the Kautxuck with Indians, belonging to the Inca and Guaranies tribes, who live on high mountains of the Andes and the plains west of the Andes, an altogether different people than the Tupí Indians. The Tupí Indians are a lively, swift acting people and the Incas and Guaranies are a dull indifferent sort of people.

They are kept slaves by the Peruvians and are bought and sold, whole families and single, according to how they come. The selling and buying is done in the following manner. If a man wants to buy an Indian, he goes to the seller and looks over the stock. If he finds what he wants, he will ask the seller how much the Indian owes him. If they agree to the price the buyer will pay the Indian's debt and the Indian belongs to him, and compels the Indian to work to pay off the debt, until he in turn sells him the same way to the next one, for whom they work from morning till night giving them nothing in return but a very scant provision such as a box of sardines and a little farinha once in a while, and hardly any ammunition. As the Indians go on through the jungles they set traps here and there catching small prey for their support. As the kautxuck is worked entirely different than the rubber (boracha) they have no permanent staying place, but are continuously moving on through the jungles, and therefore are unable to plant anything, or lay in stocks of live turtles, honey or nuts; they are compelled to look for their needs each day. One good thing for them is that they do not raise many children.

The kautxuck is worked an altogether different way than the rubber. When the rubber tree is bled or tapped each day they gather the sap. The

---

[1] Boracha is the Portuguese for the true rubber extracted from the rubber tree, *Hevea brasiliensis* in the Euphorbiaceae or Spurge Family. Caucho (spelled kautxuck by our writer) is from a member of the Fig Family *Castilla ulei* which gives an abundant white latex that is also used for rubber.

Indian boy with a basket of beijú.

The writer cut down caucho trees to get latex. This is another tree which must be cut to obtain its rubber-like latex, the massarunduba.

kautxuck tree is cut down or rather has to be to get the results, of course, only trees from a certain size up, say about eighteen inches in diameter. The kautxuck tree grows very big, much bigger than the rubber tree, and grows very plentiful. It only pays to work the kautxuck in big gangs, so the Swiss and myself picked among those that were willing to go with us, and which we thought the best, we picked twenty. We set out going to a certain point along the river, a place that everyone can well remember; each one was given a direction to go. We started off spreading out fan like, armed only with a Winchester rifle, and plenty of ammunition and facão to cut the way through the dense brush carrying a hammock to sleep in, an axe, and two dozen pieces of soap and a fire stone to make a camp fire.

On the way you cut down every kautxuck tree that is met, which is done in the following way: the brush, bamboo and vines around the tree are all cut out and cleaned away to allow the swinging of the axe which is done very carefully; for if a man cutting down a tree gets caught with an axe he might give himself a very bad cut and many a man has perished because there would be no one there to help him.

Thus preparing the tree it is bled about four feet from the ground which is done by taking the facão and cutting a groove clear through the bark about two inches wide all around the tree. By doing this the sap a foot from the cut upward and about six inches from the cut downward, will run out forming a ring right in and a little below the cut, which the next day is peeled off and rolled into a ball. Bleeding the tree this way is done to allow the cutting down of the tree with the axe, otherwise with every stroke of the axe the sap would fly into the face and hair, beard and eyebrows, and it is very hard to get off. Having bled the first tree he goes on to the next tree and so on for that day. If a man has a little push he can bleed from about fifteen to eighteen trees in one day. The next day he comes back to the first tree and cuts it down, which will take him, according to the size of the tree, from one to three hours. Although there are kautxuck trees that would take from one to two days cutting them down, that is too much of a job for one man considering the ill health, so they are generally passed by, but a big tree like that would outpay three to four smaller ones. Immediately after the tree is cut down the man sets to work cleaning little spaces on the ground, or if possible under the *kauazu* leaves. Having done

A tapir, the largest mammal in the Amazon forest.

A young peccary or wild pig.

this he cuts grooves into the bark in the same way he prepared the tree. While the sap is running into these holes, which will take about an hour, he prepares another hole about four feet square and about six inches deep, he then collects the sap from the small holes into the big holes. Having done this soap water is made in a calabash shell or a bamboo, poured into the sap and stirred with the hand until it stiffens up about the same as rubber. There is a very peculiar thing about the kautxuck sap, although it is put into a hole in the ground and stirred up it will not mix with the ground or pebbles, it stays perfectly clean.

As soon as it is set up the kautxuck is lifted out of the hole and left there, the men going on to the next tree repeating the same thing over again. As I mentioned before we spread out fan like, of course, the first two weeks we were able to keep in touch but after that spreading out more and more we lost track of each other, it raining every day and all by yourself. Although I carried an unlimited quantity of salt. I for my part always managed to have plenty to eat, game there is plenty. There is the *tapir,*[2] he is about the size of a mule not quite so high, but considerably more fleshy, they always travel in pairs, and furnish a delicious piece of steak, it takes a good shot to get one, for although they are kind of clumsy looking, they are very swift in moving about when they are followed. It is advisable not to shoot at them unless it is a sure shot, they are the biggest animal in the Amazonas. Next comes the puma called *onça,* a species of the tiger, he stands about three feet six inches high; they also travel in pairs and are very powerful and sneaky. He will follow us up for hours always keeping within a certain distance and always keeping us busy looking out for him. We have to keep a fire burning all night as it is the only way to keep him away, they are very dangerous in breeding time.

There is the jaguar, another sneaky beast and a worthy brother of the onça, he is called *Gato lepro;* we fear him more than the onça. There is the ever present *jacare* (alligator). There are three kinds of pigs called *porco,* there is a great big one called *porcão,* they travel in pairs, and are very lively. They will attack on sight, either you have to be swift and a

---

[2] *Tapirus terrestris,* the tapir is the largest mammal in the Amazon forest. In many areas it is quite rare today from over-hunting.

A pair of guans.

A toucan.

sure shot or climb a tree. Then there is the *queixada* who travel in enormous numbers, it is impossible to guess their number; they are a peculiar bunch for as soon as they scent danger they work their jaws like a scissor, rubbing their teeth and being such an enormous number they cause an ear bursting noise. A gun is of no use here, the only way of saving your skin is to get out of their way. There is a very small one called the *porquinho,* who does not get any heavier than perhaps an hundred pounds, they travel in pairs and are harmless, they are all very tasty. There are two kinds of deer, a small one red and white spots called *capoeiro* and a larger one pale red also very good eating and very plentiful called *viado.* There is a water hog called *capivara,* who feeds on grass along the river, very delicious eating; they travel in bunches and it takes a man with good nerves and a sure shot to hunt them alone. They get furious when disturbed and go right for you. Although they are as fat as a pig they can jump like a cat. As a rule they jump too high in their mood so when hunting them we keep close to the water edge and dodge them when they jump. They splash in the water snorting and barking making enough noise to scare any army. Before they can find their bearings and come back out of the water one

The Amazon land tortoise *(Geochelone).*

can make his get away. Coming out of the water and seeing no more of the enemy they make for their shot companion and if he didn't get killed the others in their rage kill him, causing a terrible commotion. The hunter who made his get away had hustled a distance down the river and now shoots, the water hog hearing this in their rage make towards where the shot fell, while the hunter makes back towards where he shot the animal, cutting out what ever he can carry and to get out of harm's way. It is the best hunting of all, for it causes no end of excitement. There is a great amount of small game, I only know them by their native names, I will name a few of the most liked ones. *Paca* is the best of the small game; *taxtu,* the *cutia* they both live on what we call the Brazil nut and are therefore very tasty; the *corocoro,* he is not only good eating but he gives us the time twice a day, at six o'clock in the evening and again at eight o'clock in the evening, calling for about four or five minutes. The *unoure,* not always eatable on account of eating certain fruits which make meat bitter.[3]

There is plenty of fowl: the *mutum,* is about the size of a turkey, black with a red knob on his beak; there are two or three kinds of a bird called *gnambu,* the largest one is called *gnambu raba,* he is about the size of a big chicken, fine eating and like a chicken lays plenty of eggs, pink of color. That one is a little smaller and is called *gnambu relayo* this is a valuable bird and although she is fine eating, she is very little hunted, she like the coro-coro gives us the time at seven o'clock in the morning and again at five o'clock in the afternoon and at eleven at night. There is a fine bird called the *yaca* but only a skilled hunter is able to bring one home. There are lots of smaller birds that we eat such as the *Qducan,* the *Papagaio* parrot of which there are different kinds, the *uru-uru* and the *pato.* To name them all would be too much.[4]

Then there is the land turtle called *yabuti*[5] they are of a pale yellow color and grow very big. They furnish a fine dish. The rivers are alive

---

[3] See *Notes* at the end of this chapter.

[4] See *Notes* at the end of this chapter.

[5] The jabutí or land tortoise, *Geochelone,* is a common animal in the forest, about which the Indians have many legends.

One of the Amazon turtles.

with fish and turtles. We go early in the morning in the canoe to the next beach and it does not take much for a man to turn over five or six turtles and then bring them home one by one and put them in a pen, feeding them on fruits and butchering when needed. They are called *tatarga,* they lay plenty of eggs, tasting every bit as good as those from the chicken. We keep good watch on them to catch the young ones when they hatch out. We then catch them and bake them the same as we do the soft shell crabs. A turtle lays from two to three hundred eggs so there is plenty of them. When the turtles make for the water the big fish are waiting for them, so is the alligator; but still enough escape to keep the river alive with them.[6]

Having given you a fair description of the animal life in those jungles, I will show the other side also. In the rainy season the jungle is alive with all these animals and the fever does not bother us so much and if it does only in a slight way. As soon as the dry season sets in things start to change,

---

[6] There are several species of turtle in Amazonia, the largest of which *Podochnemis expansa* is threatened with extinction because of over-use of both the eggs and the meat.

fever attacks come more often and more severe, the jungle is quieting down, day after day one begins to miss the different birds and animals, until the jungle has completely died down. After the fifteenth of April there is no more rain except an occasional thunderstorm. By the end of May there is nothing left but flies and wasps, even the mosquitoes are getting scarce, no more mosquito nets are necessary. It is useless to go hunting so we have to fall back on preserves most of the time. During this time anything that comes across the rifle is welcome from a monkey to a parrot.

Animal life in the jungle moves back and forth like the tide of an ocean, when they start to move birds and animals move alike. It is the birds that one misses the most, there are no singers amongst the birds of the Amazonas, but there are some beautiful whistlers. I often wondered where the countless humming birds went to, which would hang like so many diamonds above your head. Where did the *mae da lua* (mother of the moon) go, who would wake you up nights with her sentimental call, that long tailed bird would follow you up all day long calling tirelessly *bem te vi*[7] which means "I see you", but sentiment has no place in our heart, there is a hard task staring us in the face and we must pay our debts in order to get away from this mess.

---

[7] *Mae da lua* is the name for the long-tailed potoo *(Nyctibius aethereus),* a nocturnal bird. *Bem-te-vi* is the great kiskadee *(Pitangus sulphuratus)* with the distinct melodious call that is put into words in many languages, as bem-te-vi in Portuguese ("well I see you").

Notes

The animals referred to are the following:

Onça or jaguar is the *Panthera onça*.

Jacare or caiman is the *Caiman crocodilus*.

Porco or collared peccary is the *Tayassu tajacu*.

Queixada or white-lipped peccary is the *Tayassu albirostris*

Porquinho — since there are only two species of peccary, our writer is probably referring to the young of the collared peccary.

Capivar or *Capybara* is the *Hydrochoerus*.

Viado or white-tailed deer is the *Odocoileus virginianus,* the same species as in North America.

Capoeiro or brocket deer is the *Mazama*.

Paca is the *Cuniculus paca*.

Tatú or armadillo is the *Dasypus novemcinctus*.

Cutia or agoutí is the *Dasyprocta*.

Coro-coro is the *Dactylomys*.

Unoure is unknown to me.

The birds referred to are:

Mutúm is the curassow.

Gnambú, usually written *Inhambú,* refers to the tinamous.

Gnambú relayo is probably the Inhambú-relogio, *Crypturellus strigulosus,* one of the tinamous.

Coro-Coro is the name for the tara *(Phimosus infuscatus).*

Yaca is unknown.

Qducan probably refers to the toucan.

Papagaio is Portuguese for parrot and could be any of many different species.

Uru uru — the name uru is generally given to the wood quail *(Odontophorus gujanensis).*

Pato is the generic term for ducks.

9

When we set out to work the kautxuck we knew that before long we would lose track of each other so we agreed that we should keep on working until the *piquiva*[1] fruit should fall, which is around Christmas time, then we would not all start back on the same day but we counted that would take place within two or three weeks of each other, the first one to start camping until the last one would arrive. The idea now was to collect and get all our kautxuck that we made to the starting point, so that we could float it down the river. Each one had to attend to his own route, a task beyond all expression, in the first place we had traveled many miles. The average kautxuck cake weighs two hundred and twenty-five pounds. I myself had thirty-three cakes, which was over seven thousand pounds. To carry them on your back is impossible, the only way to get them out was to tie a vine on to the cake and drag them out. I did this dragging them always back to the next tree, this way I did not have to drag them such a long distance at the time and I could rest while walking slowly back to the other tree for the next cake, and besides I had them more together. It was a killing job and I swore then and there, never again, even if I had to stay there a thousand years.

Three men did not return so we had to set out and look for them, that is, if it was possible to learn their fate and bring out their kautxuck. Following their route we found the remains of one and no signs of the other two. However, we brought out their kautxuck and floated it down the Boca de Xapurí, cleaning up five thousand mil-réis each man. This went a good ways towards paying my debts. As the steamer was expected within about three weeks, we had that much time to rest ourselves. Here I made the acquaintance of a nice young man from Philadelphia; he also was there in his fourth year the same as the Swiss. The two of them had met before and were good friends, so after that we three always kept together. It was their advice not to bother with any more companions, so when the steamer came and we had got our outfit, I went alone, however, the three of us decided to stay within a couple of days journey from each

---

[1] Piquiva is the *Caryocar* sp., usually called piqui.

other so that we might keep in touch. To the disappointment of the agents the three of us did not take one third of our regular outfit. We had learned a lesson working in the kautxuck, instead of having a five thousand mil-réis outfit, we had a little more than a thousand. For instance, we cut off coffee, sugar, tea, farinha, all preserves and liquor, except Guinness stout, which we take for physic, mixing it with salt. In fact we cut out anything that was possible and that we could do without it. I was a little shakey at first to go alone, but once on the way I soon found that it was the best way at that.

I had a small canoe going up the river which I could handle very nicely. The three of us always camped together until we got at the parting point. I did not go back to where I had been previously, but went on the other side of the river and built my house about three days journey inwards. From here I could reach the maloca easier; in two days I could reach the Swiss and in four days I could reach my Philadelphia friend. Being alone I was more careful, I did not have to worry about being a nurse or a grave digger. I have had the experience of working the trees at night so I worked them as often as I could, being in fair health, for I got to be more

The flowers of the vanilla orchid.

acclimated. I had a very good harvest besides cutting down expenses to one-third of what I had used the previous seasons. Instead of coffee there were other things. Cocoa in the first place grows abundantly all through the jungle. Cultivated cocoa never comes up to the standard of the cocoa that is left to its natural state. It is very easy to prepare. The seeds get dried and roasted, and then ground up between two stones. The mash is then taken and pressed into round bulbs about the same way as one makes a snow ball; the next day it is as hard as a rock, grating it up on a tin can when ever wanted. Of course the oil is in the cocoa, home made that way. Then there is the sasafras[2] tree, the leaves make a delicious tea. Cinnamon[3] grows everywhere, the bark we do not bother about, but we use the leaves making a nice cup of tea.

Then there is a species of a grass, it grows in bushes and the leaves are white and green striped, it has sort of a lemon taste called *capim santa.*[4] Next there is the vanilla[5] which we use here for flavoring; it grows in abundance everywhere, the blossoms make a lovely tea, we drink that the most for one never gets tired of it. When the vanilla are black and white they are no good, but when they are green they make a good healthy dish. They are about the same as string beans only but they are more tender than the beans. There is a long silver haired animal who hangs on the branches of the tree feeding the young leaves, weighing perhaps fifty to sixty pounds. The meat of the animal which is called *preguica,* [6] tastes like

---

[2] There is no true sassafras native to the Amazon, but the leaves of many other species of the same family, Lauraceae make a delicious and refreshing tea. Perhaps the reference here is to *Nectandra cymbarum* known locally as pau sassafraz in Amazonia.

[3] True cinnamon is an Asiatic plant, however, cinnamon as used here refers to another member of the laurel family, *Aniba canelilla* which has a highly aromatic bark and is much used in local medicines.

[4] This is *cymbopogon,* a cultivated rather than a wild species in the Amazon forest. Many groups of Indians cultivate it.

[5] Vanilla comes from the seed pods of an orchid which is a vine and is native to the Amazon forest.

[6] The preguica is the sloth, *Bradypus,* a common Amazon animal. It lives in the *Cecropia* tree and eats the young leaves. This is nothing to do with the vanilla orchid and so there appears to be some confusion here. However, the hanging fruit of the *Cecropia* superficially resembles a vanilla pod.

A frontal view of a sloth.

mutton; this we cook together with the vanilla beans, this makes some dish. Of course we can not carry cooking pots, so we use kerosene cans; as a rule each man gets three cans of kerosene, six gallon cans. We use that for making fire and frying dried meat. When a man gets through with his work around half past nine and has to get up and on his way about half past two or three, he does not always feel like cooking, so as soon as we are through with our work we take a good bath and then hungry from the long tramp we quickly get out a piece of carne seca (dried meat), stick a rod through it, dip it into kerosene and burn or let it burn off taking care not to let the greasy parts take fire. As soon as the kerosene is burned off we scrape it a little with a knife, get a guite with farinha and the supper is ready. The first couple of times this does not taste so good, but after a while one gets used to it, then it is all right. Of course it is not a fancy dish. Very often we shoot some game on our way, but are too tired to bother with it and we let it go to waste, and used the kerosene.

Instead of sugar there is plenty of honey, there are four different kinds and the best is most plentiful, but the most risky one to get, but there is a lot of sport attached to it. The bees are about an inch and a quarter long, soot black, and have the shape of wasp. They are fighters and many a one has paid with his life by attacking them without any precautions. Running away is of no use as you cannot get through the bush fast enough for them and they will follow you for miles, even jumping into the water, the results are that you will drown for every time you stick your head out of the water to breathe they get at you closing up your eyes in no time. They build enormous big hives amongst the branches of the big trees, of black wax. To get some of their honey, I made myself a rubber blanket, six by twelve. I then went and cleared a space big enough so that the blanket would lay flat on the ground, I then put some dirt on three sides. Having done that I took the big basin that I used for making the rubber and put it under the hive. I put a cartridge in the rifle and loaded the barrel with small pebbles. I then fired into the bee hive, threw the rifle aside and dove under the rubber blanket drawing the uncovered side in tight to make sure that no bees would get near me. I could hear the honey dripping into the basin, but I could also hear thousands of bees buzzing on top of the blanket enraged. Of course I had waited doing this until it was dark, knowing that the bees could not travel down. Waiting a while to make sure that none

An açaí palm *(Euterpe)*.

of them was left I got from under the blanket and found the honey all over the ground and still dripping, the basin was too heavy to carry so I dumped half of it out and put it on my head, grabbed my rifle and made my way home. In the afternoon of the next day I sneaked around to see the damage and if possible to get the rubber blanket. I found the bees busy patching up their house and carrying the honey up from the ground. I did not dare venture near the blanket so I let it lay a few days, then I went there in the evening and did the same thing over again, I repeated this several times as I found out it was a good way to get the honey. I showed the Indians my method of getting this valuable honey without danger, which served to make our friendship more closer. The Indians bothered very little about this honey on account of the danger connected with it, although they liked it very much. After showing them my way of making the rubber blanket they certainly made use of it.

As a rule each man gets five bushels of farinha at one hundred and twenty-five mil-réis a bushel, this makes quite an item. This we cut down to one bushel using instead the more nourishing *palma*. This is taken from the açai palm,[7] which also gives a highly nourishing fruit as I had already mentioned in one of the foregoing chapters.

To get this palma, the palm tree is cut down, the crown cut off and opened, and the heart taken out. This is about two feet long and three inches wide, very tender and tastes exactly like nice fresh walnut. It is eaten either raw or boiled. Then there is an abundance of *castanha*[8] which is known here as Brazil nuts which we also eat either raw or boiled or roasted, which we use to cut out preserves such as sardines, corned beef, mutton, chicken, salmon and shrimp. This was only common sense besides paying from twelve to eighteen mil-réis a can. It does not keep, in about three months it all blows up like a balloon. Every time one would stick the point of the facão into the can to open it the air would blow out like as if there was compressed air kept in it. The only way to eat the contents was to put enough pepper and salt on it, until you could taste and smell nothing but the pepper. The condensed milk would dry up and get so tough that even hot water would not melt it anymore. There is a tree called the *canjarana*,[9] the lumber from this tree is like mahogany, this tree we tap and it gives a very tasty milk, one cannot drink it right from the tree, as it is too fat and rich.

It has to be diluted with water, each man would get three to five gallons of *cachaca* (cane brandy), at three hundred and sixty mil-réis a gallon, for it is impossible to be without it in case of a snake bite or a scorpion bite, besides after having a bath before going to sleep at night it takes away some of the effects of the many mosquitoes, *piums*[10] and wasp bites. You

---

[7] The açai palm, *Euterpe* species is the best source of heart of palm or palmito. This delicacy is today a commercial product which is canned in large quantities. In addition to using wild palm trees, the açai is now cultivated to produce palm heart.

[8] The Brazil nuts, *Bertholletia excelsa* are indeed extremely nourishing and contain most of the important dietary amino acids.

[9] Canjarana is not a name which we know. The latex of several trees can be drunk as a milk substitute. The commonest is that of the sorva tree, *Couma utilis* which has a sweet potable milky latex. Various members of the fig family also yield potable latex such as amapá, *Brosimum potabile*.

[10] Pium are small black flies *(Simulidae)* which occur in large quantities along some of the Amazonian rivers.

rub the body with brandy as alcohol is not to be had. There is always a little excitement that stirs up every day, dates can not be kept track of, one just keeps working day after day, (the reader will understand that this only counts by the man who has hopes and ambition to get away from there, as most of the men have given up, and only make enough rubber to cover themselves, so that they will be able to get new provisions every year). Whenever I felt that I would like to have a Sunday or a holiday, I took one, which might happen every three or four weeks.

Perhaps twice a season I would make a trip to the maloca and stay there for a week, or sometimes making a stop to see my friends also. It was during this season that I had a peculiar encounter with a *jacaré* (alligator). I made a trip over the river to get some fish and a supply of turtles. The fish we catch we clean and smoke, taking for this a certain fish called *tambaqui,* [11] while the turtles we keep alive. The next morning I saw a big bull laying right near my canoe, where I had cleaned the fish the night before. I whipped him with a couple of bullets over his back which made him furious.

After whipping the water to a foam he settled himself to the bottom, after watching for a while and not seeing anything I thought him gone and I boarded the canoe and went fishing. I had not gone more than perhaps three hundred yards when I happened to look around and saw the big beast following me right close behind the canoe. I quickly made for the shore, grabbing the rope and wedging it in between two branches, my rifle was laying in the rear of the canoe and I was trying to make it but had no time for he had already come close up along side and started to claw the back. When we go on the river we always carry an axe with us for sometimes a tree or the honey branches of a tree fall tying up the canoe or damaging it, we can cut the way clear or repair the canoe. I had the axe laying on the bottom of the canoe, but in a little time I had managed to get hold of the axe and jumped on shore. Swinging the axe over my head I made a little side step and let the axe go with all my might into his neck. The axe went through his thick hide and I let go and jumped for my life, for what

---

[11] The tambaqui *(Colossoma macropomum)* is one of the most delicious of all Amazonian fish.

happened now is almost impossible to describe. Moving around with his enormous tail, scratching up clouds of leaves and dirt, snoring and spitting fiercely, he blindly made a ten or fifteen feet dash forward then going back and repeating the same thing over again; but the axe would not budge. He then tried to roll, but when the handle of the axe struck the ground he made a fierce jump forward crashing into anything that was in front of him, thereby increasing his fury: the axe must have gone through to his neckbone. No matter how he tried it stuck fast, I watched him until I saw my way clear to make for my rifle safely. Once I got my rifle I hit him on his back until he was exhausted, then he slowly slid back into the river. Once he got into the river his strength came back and he started whipping the water and whenever he turned over and there would be a pressure of the axe handle and the water he would jump up high falling back with a mighty splash. Soon the current got hold of him carrying him down the stream, once in a while I would shoot at him until he got far off, the handle of the axe sticking into his neck until I could see him no more. I was out two hundred and fifteen mil-réis and called myself lucky at that.

Going back to where I was making my catch, I found a nine foot rattler laying four or five feet from the fire. Raising the rifle and taking his head off was the work of an instant. I then picked up my catch and belongings and made my way homeward. Up to this time I had not attempted to make a canoe but a short time after my experience with the alligator the Swiss came and we both went to see the Philadelphia fellow to set the time for our departure as the season was drawing towards an end.

All three of us made pretty good and we stood pretty near alike. We each had four thousand pounds and we agreed to set off early and then come back and make two canoes, that is one was to stay down and receive our provision while two were to come back. The Swiss and I were to come back so we had to look for and cut down two cedar trees before we would go down. It was then while I was cutting a tree when a priest with his party led by a guide appeared at my camp to my greatest surprise. Of course the guides traced me to where I was cutting the tree and called me to come home. I was lucky enough that I had plenty to eat so that I could afford to treat them well. It was late in the afternoon when they arrived and by the time that I got home and made supper it got pretty dark when they sat down to eat. They wanted some cold water finding the water that

I had given them too warm to drink. While I was busy serving them one of the party, an Italian, without saying a word or knowing where I got my water, ran off to get some, he had not gone more than two minutes when we heard a roar and a faint yell and the priest shrieked that one of his party had gone for water. I realized what had happened and lighting a torch and grabbing my rifle was the work of a second and off I ran towards the place that I got my water. I let go of a couple of shots and swung around my torch causing a spray of molten rubber. I found the man, his head torn clear off — death had been instant. It was then I learned the real nature of the much feared guide; thinking that they had followed me I turned around to talk when I found out that I was all alone. A puma had got the man the minute he bent down to dip up the water for he was laying right on the edge of the brook with the calabash grasped in his hand. There was nothing more to be done but to be on the lookout for the rest of the night for the puma had tasted the blood but had no time to feed anything. We passed the night in prayers while the guide kept the fires high. The next day the party went down to the stream again without paying a visit to my friends.

The Amazon fresh water sting ray *(araia).*

I could not think of starting to get my rubber out to the river which was a three days journey, as long as that puma was roaming about, because I had to use the brook which is done the following way. The rubber is thrown in the brook one behind the other, then a pole is procured, wading in the brook with the pole we poke the rubber ahead whenever it gets stuck two balls at a time. After getting the first two a distance we wade back for the next two, going back and forth all the time. Once in a while where the brook is clear they will go along for a good distance. I had to get that beast for my own safety, they always travel in pairs so I had to be careful. After three days I got the male right near the spot where he killed the Italian. Having got one I lost no time in climbing a tree so that I could cover the carcass with my rifle, it didn't take very long before I heard the raking roar of the female who had scented the body of the male and also the danger, and therefore kept under cover. When complete darkness set in I could see her shining eyes as she came out sniffing around the dead body of her companion. Getting myself in a good position I made a noise which made her look up long enough for me to plant a bullet right between her shiny eyes. I could now get some sleep for I had none the night before.

Being disgusted with the place I lost no time getting my rubber on the way towards the river where I joined my two friends. The Philadelphia fellow was there but the Swiss was behind and when we went to his assistance — he had been stung by an *araia* [12] which is a flat creature and lays between the rocks, it had two spores about three inches long, and if a person who wades in the water happens to step on them nine times out of ten they will get stung. It is very painful and heals very slowly. Of course we helped him getting his rubber towards the river and took good care of him; after all we made a very good trip down the river with the customary ups and downs.

Having settled our affair with the agent we exchanged things around so that the Philadelphia fellow and myself would go back and make the canoes while the Swiss stayed back, he having a sore foot. Of course the two of us in an empty canoe did not take longer than thirty days to get back and we lost no time in getting to work. In three weeks we had the

---

[12] Araia is the fresh water sting ray which is common in the rivers of the Amazonia.

first canoe in the river and at the end of six weeks we had one hid in the bushes to be used for taking our rubber down the next season and we were on our way down the river in the other one, paddling our way down the river at full speed. We made the trip downward in fifteen days still having a few weeks to rest up as the hard work was beginning to tell on us.

The construction of a dugout canoe.

The largest dugout canoe seen by the editor compares with that made by the author. This one was made by the Mayongong Indians.

10

The Brazilian government does not provide any law or orders to the men in the jungle. If anyone is caught stealing there is not much fuss made, he is stood up against the first tree and who ever feels like it can take a shot at him. This may sound harsh, but considering the conditions existing there, rubber laying around everywhere and no doors in the houses, conditions would be such that nobody would be able to do anything at all. Every seringueiro has his mark, which, when he gets a ball of rubber finished is burned in similar like marking cattle and it is a peculiar feature that it cannot get away without leaving a mark. If anyone steals a ball of rubber he is caught right off. Having been told of these conditions I had the chance of witnessing the execution of one who had been foolish enough to try and steal. It was surprising to me how quick everybody passed over such a horrible drama, for about half an hour later I did not hear a word more spoken about it. Everybody who was well or could work around was either drinking or gambling, however more than half of the men that had assembled in camp were sick and in a desperate condition. Some were rattling their teeth with chills while others were either singing, yelling and cursing with fever. Some were laying there with dropsy with their abdomens so full of water that it looked to me that they were ready to burst any minute. There were also a great many dying with beri-beri, some were laying around with *elephantis,* enormous swelling and cracking feet, there were a couple of men raving who had been rescued from the Indians with whom they had gotten into a dispute of some kind. The Indians do not kill anybody outright if they can help it, but they will get him alive and cut open his sinue at both wrists and right above above his heels, thrust about a foot long plugs through the cuts cross his arms and legs and pin him to the ground, leaving him to his fate.

Everybody is eager to shake your hand and call you *amigo* which means "my friend". I myself felt the same way, there is something in you that causes that brotherly feeling and makes you friends with everybody and everybody your friend, as long as you are square.

When the steamer arrived and we had our provisions we loaded our own canoe and each of us were in about seven hundred mil-réis, for we did not have to pay the agent for his canoes, besides taking the risk of losing the canoe and then pay what ever price he would demand. Making your own canoe is a good paying proposition, and I made up my mind to keep it up not only for my own use but for selling also. Having had a good season we decided to settle on the same side of the river going a little further up. This time we stayed closer together so that we could reach each other within a day, we would occasionally shoot bigger game and divide it up, in order that we would not have to lose so much time hunting for smaller game.

Each of us took only one bushel of farinha and when I opened mine I found it all muffy and unable to use, the agent seeing that we had been paying our debts and were cutting down expenses had shoved a leftover in on us and it fell to my lot. I decided right then to get in on him so tying up the basket well I thrust a stick through it and smoked it over carefully with rubber, making a handsome ball of it. This being my fourth season I expected to come out with a good margin and then make my getaway. I concentrated all my will power towards this, instead of working three estradas I opened up five working only at night, five nights a week, one night on each estrada, the other two days I would hunt or sleep or work on a canoe, and on one night I would yield more sap than three days and I got ahead with leaps and bounds. When my two friends paid me a visit they could not see how I did it, and explaining to them my system they followed up the same and soon had things humming. It was then that we made a pact between us that if any harm should come to one the remaining ones or one should own what was left. Everything went along smoothly except an attack of fever. As usual, we were pestered an awful lot by the constrictor who visited us every night; they make you nervous especially when one wants to get up and get a drink of water you have to disturb them which makes them angry besides the weight of their head and neck on your chest gives you the nightmares. The neighborhood seemed to appeal to the leopard[1] for they appeared very numerous. We had an almost daily encounter with them and I brought quite a few down and tried to save that beautiful skin. I dried the skin in the sun until it was as hard as a board

---

[1] This must refer to the jaguar, since there are no leopards in the New World.

but soon as I would roll them up and hang them away the flies would get at them and the next day they would be full of worms, I gave up. I caught many beautiful butterflies and tried to save them but with the same results.

Between cutting rubber and making a canoe time passed along very quickly and before we knew the six months had passed by and when we made the Boca do Xapurí by the middle of January I weighed off with four thousand kilos of rubber (eight thousand pounds) at seven mil-réis a kilo, making twenty-eight thousand mil-réis, a nice little sum and with the seven thousand I had made the season before I had thirty five thousand mil-réis. Besides what I got for the canoes which paid for the provision that I had gotten it left me enough to pay my way down without touching the gross sum. Here is where I made the biggest mistake I could ever make. Instead of embarking quietly, I went around and told some of the fellows that I was going down. Soon I had everybody around me, especially the agents, patting me on the shoulder and telling me to stay and triple the money and make it worthwhile. I gave in after a while and let the steamer go off, figuring myself that I would never have a chance to make that much money again and here is where I lost out. The three of us decided to stay together

A storm over the river Acre.

again but the two did not want to go up the river Acre, they wanted to go up the river Xapurí, that is the river that Ex-President Roosevelt styled the River of Doubt, a river of not more than a hundred to a hundred and twenty feet wide but very deep and a swift current, which makes it very hard to paddle the canoe upstream. The valley through which this river flows had a terrific under growth, almost impossible in some places to get through. After a three days journey I did not care to go any further and I made a raft to go back, leaving my whole outfit behind which I renewed when I got back at the Boca do Xapurí got a canoe and went up the river to my old neighborhood. I made a little over a thousand kilos that season for I was laid up with the fever most of the time. I left early that season and when I arrived at the Boca do Xapurí I learned that my two friends after traveling ten days and going through the hardships returned and were a little way up the river working some old estradas, they had lost too much time traveling on the Xapurí River. I did not wait long in camp but went into the jungle gathering *copaiba oil*[2] which is a very valuable oil for medical purposes. It is taken from the copaiba tree, a tree that is very scarce and not every tree gives oil.

I spent all the rainy season gathering five gallons, an experience similar to the kautxuck work. Of course, the copaiba tree does not get cut down; a long V-shape cut is made clear through the bark and then a cup is pressed into the bark and the whole thing covered with *causu* leaves so that the rain will not touch it or get into it. You have to tap about a hundred trees, going to and fro through the jungle, watching all the time for leopards, pumas and snakes; also keeping your mind strained not to get lost and to find again every tree that you have tapped. You carry a hammock to sleep in, a rifle and a rubber bag strapped on your back to put the oil in. Out of the hundred trees that are tapped, perhaps fifty will yield a little oil. Having gone over the tapped trees and collected the oil, another hundred or so are tapped and so on. You have nothing else to eat but nuts and small game. A bath once in a while during the day; a comb and a brush and a

---

[2] This is the oil of the copaiba tree *(Copaifera sp.)* a member of the Bean Family whose trunk produces the oil which is combustible and will run a diesel motor. It is also much used by the Indians for its medicinal properties.

towel are unknown things. When night comes you look for a safe place to hang your hammock up to sleep high enough so that the puma can not scent you.

When the time came near for the steamer to arrive I returned to camp. As soon as the steamer arrived I took my oil to the captain as he had bartered on previous trips to collect some oil for him, offering a big price. Showing him the oil he took me into his cabin to pay me for the oil; he opened a box, took out two cotton shirts which can be bought here in any dry goods store for a dollar. He told me to pick out the one I liked best, at first I thought it was only a joke, but when he insisted I asked him what he meant, he coldly told me that he thought I was well paid. I then completely lost control of myself — he was a man about fifty and no match for me. After letting my temper out on him, I put my knee on his throat forcing him to open his already bleeding mouth, I poured some of the oil down his throat until he was choking; I left that scoundrel in a prostrate condition and threw the rest of the oil overboard.

The bung in the trunk of a copaiba tree. When removed the oil runs out. It was this oil which led our writer into a dispute with the riverboat captain.

I made my way quickly into the camp as I knew once I was in camp I had every man in back of me. They seeing my excited condition I was surrounded and asked what had happened. Telling them my experience the whole camp was soon in an uproar. After a while the first and second mates with some of the crew came into camp and demanded that I be turned over to them. That settled the case, the Swiss who had the most influence could not even handle them anymore, the whole camp tore loose. The mates were soon overpowered with their men and tied two and two together, then the whole bunch pressed forward towards the steamer and began shooting the steamer up along the water line. The captain knowing the consequence was smart enough not to answer the fire; the boat was made useless and the captain with his crew who had received a good shaking up had to go down the river in an open canoe and what was worse no rubber went down. The whole thing was a frame-up between the captain and the agent, who were trying to cheat me out of the oil. The agent let the cat out of the bag as he told one of the men that I had cheated him by smoking a bushel of farinha over with rubber, which the captain had brought back. They knew who did it by tracing the mark. He also told the man he would get square on me. I had turned over the rubber but had not received any money. When I heard about this I took my rifle and asked the Swiss to come with me to see the agent. When we arrived there the agent was busy dealing with some of the men, waiting for my turn I asked him for the money. He counted the money out on the counter telling me that he was to discount for the ball that had to be returned. There was no use arguing so before he knew I leveled my rifle at him and told him to hand the money over to the Swiss, giving him two minutes. He hesitated but seeing my gun was not the only one he turned over the money. I took the money for the ball of rubber that had the farinha in it, the rest I returned. I took my rubber back and loaded it into a big canoe together with my provisions and went up the river for the sixth time or season. This time I only went thirty days up the river to a place that I had selected while I was gathering oil. It was one of those rare places where the rubber trees grow close together; working five estradas by night I was able to work two hundred to two hundred and twenty-five trees on each estrada. I made a big haul that season although I suffered a lot with the fever, for the place was a fever hole and alive with everything that was bad.

Here I had a narrow escape of being torn to pieces by a puma. One of my estradas went zigzagging along with a little brook for a few miles; there was a good size rubber tree growing slant over the brook; in order to tap the tree I had to stand my rifle aside and hold on with one hand and bend down to tap. While I was doing this one day I saw a puma standing about five feet away, his big round eyes nailed on me, he then let loose that bone raking jaw. To get my rifle meant death there was no time for thinking so I let myself drop on my back into the brook causing a big splash. It saved my life as that unexpected splash gave him a fright and he disappeared into the jungle. Once I had my rifle I felt safe and continued on my work.

I had bought some watermelon seed, muskmelon seed and peanut seeds. In the first chapter I mentioned that the Amazonas River with its countless tributaries rises six months and falls six months. When the river falls big sandy beaches appear, they are very fertile and anything planted will grow in abundance, say; tomatoes, beans, watermelons, muskmelons, peanuts, sweet potatoes, sugar cane and different kinds of tropical vegetables. The men seeing me buying those seeds, knew the time when they were ripe and would come floating down the river, too lazy to plant any themselves, and watch for them floating by. The Indians took a great liking for them also and I would think nothing of it to throw forty or fifty melons into the river and let them float down for somebody. In course of time they had got used to it and came to demand it from me, although they had just as much chance planting for themselves.

I found a monstrous big cedar tree, the biggest I had seen yet, so when the season came to an end I decided to take enough rubber down to buy provision for myself and then come right back and make a canoe out of this cedar tree.[3]

Having made a good season of rubber I cut the season short and built a scaffold around the tree high enough to where I could cut it down. The tree was so large that it took me three weeks steady work before I had it laying on the ground and another week to cut off the length which was

---

[3] The Amazon cedar tree *(Cedrela sp.)* is a member of the Mahogony Family that is the source of one of the best commercial woods of the region. Cedar wood is still much used for canoes and boat building.

sixty-four feet. Having accomplished this I got ready for my trip down the river, I did not get started for I was taken sick with the yellow fever and although it was only a slight attack it took me four months to recover, being alone. I would never get over it if it had not been for the Indians who came and looked after me. This yellow fever is a terrible thing, it starts very suddenly with an unquenchable thirst, followed with pains in the back and legs and a high fever, and as the sickness progresses, violent headaches set in and the eyes and fingernails turn yellow. The Indians would make tea from certain herbs which they would boil in a calabash shell. Of course, they could not push the calabash shell on the fire, they would put suitable stones in the fire until they were red hot. Having put water and herbs in the calabash shell, they would put one stone after another stone into the shell until the herbs were well boiled. The tea thus made looked exactly like ink and took me months afterwards to get rid of the stains on my lips and hands, but it did me good. After I had recovered and was able to go about my work it had become too late to go down for provisions so I made up my mind to stay up there altogether and make the best of it, relying on my gun and turtles and what I could catch in traps.

Yanomami Indians, in their hammocks.

11

Having decided to stay up for my seventh season, I spent the next two months carving the canoe into shape which was quite a task. I had tons of wood to handle and when the Indians came and saw what I intended to do they shook their heads and gazed at it. When they came and saw me start to carve a canoe out of that monster block, they sat down silent watching me, once in a while making a remark. They would come every day watching me progress in my work, always willing to tumble the canoe over when necessary. I had quite some coffee left and they took advantage of it and kept on making coffee now and then during the day. I did not bother much with them as I was too busy getting the bark from that canoe, which was from three-quarter to an inch thick, into shape. I soon realized that the job was too much for one man. I knew also that if I threw the job up the Indians would sneer at me and I would lose their confidence forever, therefore I had to grind my teeth together and keep at it, until it was time to start at the rubber again.

Six months were staring me in the face with no provision, and not even enough ammunition. I decided to go a ways down the river and meet some of the boys coming up and get whatever ammunition I could get besides a little news.

On my way down the river I came near losing my life which happened the following way. I ran into a shoal of pitapitinga, a fish which I have already described in foregoing chapters, as one of the most looked for fish in the Amazonas river. I naturally decided to get some. They run along from about ten to twelve pounds. In order to keep them a couple of days we clean them, leaving the scales on them. We got out enough fresh sticks to make sort of a grating about two feet from the ground; the fish are sliced into halves and laid on top of the grating under which we keep a low fire taking care that there is plenty of smoke so that the fish gets baked and smoked at the same time. When I had caught a couple of them and went on right down the river, looking for a good place to bake them. Sitting as customary in the front of the canoe, paddling a couple of strokes to keep the canoe straight I did not bother in looking in back of me. All of a sudden I heard a terrific roar and at the same time the front of the canoe shot up in the air, throwing me headlong with paddle and rifle which was laying across my knee into the river. A great big bull of an alligator drawn by a smell of cleaning the fish had followed me up, while I was merely drifting with the current, it had grabbed the rear end with his claws pulling the canoe down with his mighty weight. When I got to the surface I found myself almost alongside of him. It was now a race for life to keep away from his claws and the mighty strokes of his tail. He was pounding the water into foam with his tail trying to put me out of commission, but in doing that the brute put himself out of commission and that proved my salvage.

I now found myself in the worst predicament that I had ever been in since I was up in the Amazonas. In the first place I completely lost my nerves, then my rifle was gone also my facão; and I did not know the exact spot they had sunk, and even if I had known the spot I would not have had the nerve to dive for it. Paddle and canoe were drifting down the river. Being without any arms whatsoever the first thing to do was to make a fire for the night and hug close to it. I experienced quite a strange feeling of being without arms, all night long I was continuously hearing or seeing something, and under such circumstances there was no chance of getting any sleep. The only way to get out of that mess was to try and get hold of the canoe and make my way down the river as intended and to live on fruits that I could find. On the other hand I knew I would perish one way or the other; so when day came I lost no time thinking, but started to elbow my way down along the river, only to find out that without a facão it is impossible for along the long river edge, the undergrowth is the heaviest and mostly thorns. I gave it up for a while and ventured a little further into the jungle to find some nuts that I could have something to eat and calm down a little. Having lost pipe and tobacco the mosquitoes almost eating me up alive.

Along the river and swamps there grows a tree which is called *pau bomba*,[1] because the stem of this tree is full of hollow chambers that never get more than six inches thick. When they are about three inches thick a man can break them off fairly well. After having something to eat I set to work bending down the bushes to make an opening towards the river edge.

---

[1] Pau bomba is not known to me.

I then went to work bending off these young trees so that I could make a float binding them together with vines that grow everywhere. That took me all day, I made two good fires this time but enough apart so that I could lay safely in between and get some sleep. I covered myself up with palm leaves. The next morning I set off with my raft down the river using a dry limb as a pole to guide the raft and to push ahead and catch up with the canoe, which now had two nights and a day headway on me.

I soon had my hands full of blisters from the dry limbs that I was using as a pole so that I was unable to make any headway at all. All that I could do now was let myself drift down with the current. My only hope was that the canoe would get caught along the river somewhere. Nine days I drifted down the river when I caught sight of the canoe turned upside down in the bushes. It did not take me very long to turn it over and get the water out and I then continued on down the river using small green branches to guide the canoe, which I would break off along the river when necessary.

Late in the evening of the sixth day I ran into a camping party at the mouth of a small river. There were four in the party, all sick and disgusted. They had a little fire smoking and were sitting about, not even enough ambition to make something to eat. They were just as astonished as I was glad to see them. Although I had been living on fruits for sixteen days I was in fair condition. After the customary greeting — they were all strangers to me — I got busy making a good fire for the night. I borrowed a rifle and went into the jungle to get something to eat; I did not have much time as night was setting in. After crawling around on my knees for about fifteen minutes I succeeded in getting a *tatu*, a very tasty and fleshy little animal, known also by the name of armadillo; it had no skin only a shell over the back, tail and head. Well, we had a good supper. We stayed two days at the camp and they only having a rifle for themselves I had to keep traveling down the river, while in turn they went up the river towards their destination. Two days later I met two canoes coming up, men that I knew. I got a sixteen shot rifle and learned the sad news that my Swiss friend had died from beri-beri. They wanted me to stay with them but I had made up my mind to finish the canoe and besides I wanted to be near the maloca where I was befriended by the Indians.

When I got back to the place where the alligator upset my canoe but I could find no trace of the big brute. I kept on until I reached my old place, finding everything the way I left it. I only needed a couple of days rest to put me right to work at the rubber. I was in for a peck of trouble however. The Indians scouting around the jungle had noticed my returning and lost no time in paying me a visit which consisted merely giving me a nod and sitting down and waiting for a cup of coffee, speaking only when I would speak to them. Nevertheless they gave me valuable advice. Of course I was glad they would come, besides they would always bring along plenty to eat, either some rare fruit or some game that they knew I liked; and whenever they came direct from their maloca they would always bring a bundle of beiju which I have already described in one of the foregoing chapters. I told them that I was going to lay in a stock of turtles, they advised me not to go near the river for a few weeks as the Botoudous tribes were on their way for their yearly supply of fish. They invited me to come and stay a while with them at their maloca. As I had never had any trouble whatsoever with the Botoudous although they came to the river every year to lay in their supply of fish I laughed at them. Early the next day I fixed up my corral and went to the river. I succeeded in turning over twenty turtles which is done the following way. We cut a strong enough branch from the tree so as to make a fork about five feet long guiding the canoe up stream making no noise whatsoever that is the main point; then spying a shoal of turtles laying in the sand make a quick landing and with the fork when they make for the landing shove it under their breast and turn them over. One has to be very swift about it else they will all be in the water and you also have to take care of your legs as they are a bad bunch when attacked. Sometimes you get enough with one landing other times you have to make two or three landings.

Having gotten enough you put two or three according to their sizes in a canoe and bring them in the corral, or pen where you feed them with fruit and mat. They cannot lay very long on their backs, as soon as the sun warms them they die. They can stand all the heat of the tropical sun on their back as long as their chest is wet and cold. So you have to be lively and bring them to the corral. It is an exciting event. Having turned over twenty turtles which would last me forever. I loaded the canoe and made for a temporary corral at my landing place, which took perhaps twenty minutes. Returning I saw two of those filthy Moro Indians waving their hands at me although I had not seem them going for the turtles I was not the least surprised for one is apt to see them almost anytime along the shores. Thinking they wanted me to pass on to the other side of the river I did not pay any attention to them but kept right on landing at the beach where the turtles were laying. All of a sudden a spear landed right in the canoe and went through the bottom. There was no time for looking at it, I grabbed my rifle and dived into the river and made for the other side, diving as often as the rifle would permit. Outside of a big scare nothing happened to me. Everytime I stuck my head out of the water I could hear them arguing. They got hold of my canoe, but I knew they could not take the canoe into the interior, so the only thing for me to do was to hang around for a while, keeping a watch if they would set the canoe adrift. Taking good care to keep from their sight. Knowing Moro Indians were around I went hunting up very much to their dislike, if not so much as far as I was concerned. They are deadly afraid of the Tupí and Botoudous Indians, as these tribes will kill them on sight. The Moro certainly are a filthy lot although they live in the swamps and along the river. I never saw them bathing, they live mostly on alligator and turtle, and it has degenerated them so that they are full of warts and things. Catching fish is too much trouble for them, they just lay around day in and day out, so ugly and filthy that not even the flies or mosquitoes bother them.

One of the Indians had thrown their spear too soon in this attempt to get me and the others gave him a call down for my getting away. I heard them say that they would destroy the canoe and go looking for me. I took the hint and lost no time going to the maloca of the Indians, which would take me every bit of four days. Some luck. I was late for the rubber and now I had to go to the maloca and stay there for a couple of weeks until the Botoudous would be on their way to the interior; I would then start and make a canoe as the rubber season would be over. It took the heart right out of me. While I was making my way through the jungle I was cursing the Almighty and my Mother and the next minute I would be singing anything that would come to my mind, as loud as I could so that if any of the Botoudous were around I wanted them to hear me. Then again I would be cursing myself and the day I was born. I sat down time and time again and cried and then again I would laugh. I remember all this very clearly, but what I don't remember is from what I lived. I never fired a shot neither do I remember how many days that I traveled or how I passed the nights. However instead of landing at the maloca I landed at my own camping place, or the place my camp had been. The camp was all burned to the ground, there was absolutely nothing left. I was dazed for awhile and then I came back to myself and examined the place. I found that the camp had been burned a few days already. I was puzzled as to how I had gotten off of my track towards the maloca. It being about midday I thought there would be no Botoudous in the neighborhood so I sat down and cooked myself some nuts in a tin can and then set off once more towards the maloca.

After traveling for about an hour along one of the estradas I heard a very faint moan. I immediately threw myself flat on the ground, a rule that is followed in every case of danger. In the first place it puts you out of danger and then again laying on the ground one can see better and further through the under bush. After awhile I heard the same moan again, this time I was sure it was a human being. I made my way carefully to the spot where the sound came from and I beheld a sight that was awful. There was a Botoudous Indian sitting on the ground with his legs twisted around a tree in such a way that he could not free them and his arms pinned up high on the tree, having put three pins through each arm. I had heard a lot about Indians torturing their prisoners but when I saw it with my own eyes it made me sick. The poor creature must have been there some time as his whole body was covered with ants and mosquitoes and the blood on his arms was all dried up. When I pulled out the pins that held him to the tree it started bleeding again. I laid him over on his back and I twisted his legs from the tree. I stretched him out but I had to drag him aways as the ants were so numerous around him, they already eaten into the corners of his eyes so that he was unable to see who was helping him. While he was laying there I had a good chance to study the difference between the Tupí and Botoudou Indians. He was a young fellow about twenty-four years of age, of European complexion with dark hair, no disfiguring marks or holes through his lips like the Tupí Indians; neither did he have square jaws like the Tupí with a sharp nose.

After he had rested for awhile I carried him to a near brook and gave him a bath and bandaged his arms up with cauasú leaves. I now had to prepare a place for the night. I had made up my mind to stay with him until he was able to take care of himself. I could not go hunting now for fear of the Botoudous, so I got some more nuts and fruits for supper made a good fire and put the Indian as close as I could and covered him with palm leaves. I then laid down myself but could not go to sleep for the Botoudo Indian is a very treacherous Indian from what I had heard and my own experience had fully convinced me that there was a lot of truth about it. Anyway I did not care to take a chance for he might feel strong enough during the night to get up and find me asleep and go against me. I kept a good watch all night and sure enough towards morning after sleeping all night he woke up and pushed the palm leaves aside; he looked around kind of puzzled then he got up on his feet but so did I. I motioned to him to sit down he did not pay any attention to that but started to stretch his legs and arms. It must have caused him pain, judging from the face he made. Then seeing that his arms were all bandaged up he sat down again never taking his green eyes off of me. I spoke to him in Tupí language but he did not understand me. I showed him my rifle and motioned him to keep sitting down. I then got some nuts for him and then tried all kinds of ways to find out from him in what direction his people were going but could not get a word or a sound out of him. I got some fresh cauasú leaves and tried to wash and bandage his arms: but he pressed his arms so tight together so that I was unable to do it. I then picked up my rifle and left him walking backwards towards my camp a short distance; I then made a big sweep turning towards the maloca again nothing happened to me and in two days I was at the maloca.

When I arrived at the maloca there was nothing but women and children, and they plainly showed their surprise at my appearance. The men were all out and as I learned later, the greater part were fighting the Botoudous while the rest had thrown a ring around the malocca to protect the women and children. After they had given me some refreshments the women told me they knew where to meet the men and asked me to accompany them. The men were just as much surprised as the women were, they could not understand how I had gotten through. They said they had never witnessed so many Botoudous together in all their experience. They firmly thought that all the different tribes had united to wipe out the Tupí. Telling them my experience and also that I had found the captive Botoudou, they advised me not to leave them but to go back to the maloca with the women and help protect them.

When I returned to the maloca there was not much to eat, only beijú so I left the maloca and went hunting. It being late in the afternoon I made up my mind that if I did not get anything before night set in I would stay out over night; but luck was with me. After crawling about for about an hour I brought down a capivaro weighing about four hundred pounds. Packing the liver and hind quarter on my back I made my way to where we had met some of the men. When they saw me bent down with a load they forgot the Botoudous and giving them the direction they went and got the rest and delivered it to the women. They then returned to their post to wait until the Xoa-Xoa would call them in. After about two weeks the fight came to an end. The men returned to the maloca bringing in three women prisoners. They then spent a week paying tribute to the men they had lost in the fight, which was five men.

I had often asked the Xoa-Xoa to give me one of the girls, promising to marry her; but he would not do that saying that no Tupí women could marry a *kurukee* unless he would join the tribe and give up hunting rubber. That of course I would not do, so I had to stay alone and do my own cooking. After I had returned to my camping place and had built a new house, I was very much surprised one day when unexpectedly my best Tupí friend named Hico with three more men brought in one of the Botoudou women. A woman about twenty-four years old, a very pretty woman, tied with her hands to her back. They told me the Xoa-Xoa had sent them with the woman and told me I could have her. They tied her to one of the posts of the house and after having something to eat they withdrew for the night, making themselves comfortable for the night. I went and untied the poor woman and gave her some water to wash her arms where they had been tied and gave her something to eat. I then called Hico to tell her to lay down in my hangmat which I had braided of *basto* while I was at the maloca. Basto[2] is taken from a certain tree, it lacks bark and the lumber and is peeled off in long straps then pounded soft with a club and washed out and bleached in the sun, it is similar to flax. I would lay out with the others by the fire. She immediately lay down after awhile I went in to see if she was comfortable but the minute that I touched the

---

[2] Basto is the inner bark of various trees.

hangmat, she jumped at me like a wildcat clawing my face with both hands and she then disappeared into the dark of the night. My face was all scratched up. When the Indians saw me the next morning they grinned and asked me if I wanted them to go get her. I begged them to let her alone if they thought she were able to find her own people. They thought she could, so they left her go. I think it took me about four months to heal my face. From then on I worked at the sixty-four foot canoe, eating anything that I could get hold of.

It was a trying season working hard and living mostly on fruits and nuts, nevertheless I made good headway and when the rainy season set in I had the canoe high enough so that with the help of the Indians I could shove her down the river. As soon as I had her in the river I loaded it with enough rubber to get what I needed for the next season. It was a man's job to guide that big canoe single handed, but I made good headway and got it safely down the Boca do Xapurí, going through the customary hardships of course.

I caused some surprise with the big canoe, the whole camp made a festival in my honor. There were two boats unloading their merchandise when I approached one of the captains offering my canoe for sale. He sneered at me and told me it was lost work to carve a canoe that size, but, he soon changed his mind when the other captain offered me five thousand mil-réis, he offered me seven thousand mil-réis which I accepted, that was a price worth while. He packed the canoe full of rubber and took her in tow down the Manaus.

Canoes under tow by a river launch.

12

Having received seven thousand mil-réis for the canoe together with three thousand for the rubber that I had brought down, I made a big purchase of provision and in turn divided it up with some of the boys that I knew were reliable, with an agreement that they pay me back in rubber at the end of the season, I of course taking the risk that they might never come back.

I soon went up the stream again with everything in my favor. Of course it sounds very easy to say going upstream, but when the reader realizes you have forty days paddling a loaded canoe against a strong current all alone, running up against all sorts of trouble, it sounds quite different. I made my way towards my old haunt. I had been staying on the right side of the river so this time I worked my way closer to the maloca; a little too close, so I decided to switch off to the left side working more away from the Indians, lest there be hard feelings, although I did not use my gun more than was absolutely necessary.

Having established myself at a good point, I lost no time in getting ready to open up my estradas. I had settled as I found out near big lakes. I made this discovery a very odd way. On a hunting trip I had seen quite some açai palms with plenty of ripe fruits so I decided to go and get some. Making my way towards the palms I noticed that I was walking on a very peculiar ground but being up against all kinds of things I paid no further attention to it, but went on up the palm and picked the fruits. About an hour afterwards when I wanted to go back the same way that I had come I found myself looking over a lake. I went back to the palm to make sure that I had taken the right course I came back a second time and saw the lake. There was no mistake the lake had gotten and was getting bigger all the time. I was looking across a mile of water. I had never heard anybody talk or tell about anything like it before so the reader might know that I was greatly puzzled; indeed I felt very uneasy. Of course I let the fruit lay and went scouting around to find out what I was up against. I had to be very careful going about least I would step through for I was moving on false ground. As near as I can describe it was sort of a floating mass formed from grass and shrubs which in the course of time had grown so heavy that it permitted even the growth of small trees and palms. Indeed it was thickly grown with açai palms and they to my judgement were the cause of this mess moving around in the water. The wind catching in the crown of these palms would move the whole thing from one side of the lake to the other side. After this floating mass had reached the outside of the lake a distance of about two and a half to three miles and packed itself tight to the shore I was able to pick my way to solid ground. There was a long road ahead of me to my camp. In the first place I did not know the size of the lake neither did I know in what direction it was extending. There is the swamp that always surrounds the lakes in the Amazonas so I had to pick my way the best I could, taking notice of everything so that I would be able to find my way about going hunting, for it is always good hunting around lakes, there being always any amount of ducks, pelicans, capibaras, turtles, deer fish, water snakes, alligators and so forth. Taking in the sight as well as looking for good hunting places I made slow progress

A small floating island of vegetation, nothing like the large one described here. These floating islands are a quite common phenomenon as the river level rises in the flood season. They are often washed out into the main rivers where they gradually break up. They frequently contain trees.

and reached my camp the following day well in the afternoon, bringing home with me a monster of a pelican,[1] the only one that I ever shot in the ten years that I was in the jungle. Of all the game there, the pelican is the hardest to get. I brought home one that measured eight feet from wing to wing, they are very noisy one can hear them a mile off, but they fly into the big trees and rest themselves between the big branches in such a way that you cannot get them. They are very delicious eating.

There was a tremendous rice field extending from the lake all I had to do was to make a small canoe to navigate the lake to get in amongst the rice. It was very delicious rice tasting very much like our barley with a large kernel and a bluish color.

It was my eighth season and the most eventful of all, I had no fever attack at all during the whole season and of course I felt strong and able. I tapped rubber at night as I have already explained one night brings more than two days. I carved out a canoe for the lake. I wanted to explore the moving island if I may so call it, but before I had the canoe made I ran into a *grotto fundo*. The best way to explain this is that its cave split open on the top so that a scanned light penetrates into the cave. Roots of big trees have found their way through the opening at the top so that when a person looks in from the outside it gives one the shivers. I have heard others talking about these grotto fundos but it was the first one I have ever seen. Of course at first I was afraid to enter the place but after a few days I returned and looked the place over a little, but the light that came in from the top was too dim and I could not see anything inside from time to time. Finally I made up my mind to enter, but before I did so I made a big fire in front of the cave to smoke it out, for I had heard that these grotto fundos were the haunts of snakes and pumas. When I thought that I had smoked the place out well enough I entered carrying a torch of rubber so that I could see everything about me. I found no snakes or pumas but the place was alive with *morcegos,* a big species of a bat, from four to five inches high. The smoke did not seem to effect them in any way but the smoke from the rubber torch they did not like, in a minute or two the place was

---

[1] The true pelican does not come inland into Amazonia. The author probably shot a stork such as a jabiru which has a pelican-like pouch.

alive with them. Apparently they were trying to get out in the open. So I sat down awhile and watched them making their way out, but as quickly as they went out they came in again, making for the other end of the grotto fundo. I entered quite aways and seeing that it branched off in two directions I did not feel safe to go either way, so I went out and got enough wood to make a good size fire at one opening so that I could enter the other side. After going about one hundred feet or more I found the ground covered with odd stones all the same size, oval shaped about two and a half inches long and one half inch thick, black with little crystals all around the rim. I picked up some of them and took them along. I later gave them to an American mining engineer who after examining them found that they held eight percent of quicksilver. There must have been lots of gold near by. I entered the other opening only to find it in larger quantities of the stones than the other.

Having satisfied my curiosity about the grotto fundo I turned my attention towards the lake again. Getting my canoe I went into the enormous rice field, which was more interesting as there was no end to it. There must have been a thousand acres alive with all kinds of game such as capibara, coro-coro, bacú, anta (tapir), viada (deer), and tremendous big water snakes, also an immense lot of turtles. I had the time of my life hunting. There was from two to three feet of water so that I could navigate all over. How that rice grew was a riddle to me, but it was there and I ate plenty of it. It was simply grand. I killed some of the big water snakes that measured six and seven inches around the heaviest part of their bodies. They are of an orange color and very sassy. They would attack as soon as you get near to them. I had lots of sport with them. I do not know whether they were poison or not. The lake itself was alive with everything; especially with *londero,* a fish. One would see them everywhere mainly on the floating land and of course I shot quite a few of them. There were also a great many anteaters around the shores. One day, quite to my surprise, I got too near or close to a prego porcupine. All of a sudden I was full of needles, with a result that I was laid up with the chills for a few days. Before I had only met them on high ground but this one was a very large one living along the swamp shores. It was a silver gray color with needles four to five inches long. After awhile I ran into a band of Moros.

They were or acted surprised at seeing me but they did not bother me. I mingled freely with them, their only arms or spears being crudely made, but they are masters at handling them, never missing a mark.

Mingling with them I was able to study them and their customs well. They plant absolutely nothing but eat what fruit they find also nuts, spearing alligators up to about four foot lengths. They roast them over the fire and eat them with wild honey. They were very fond of anteaters and turtles, occasionally some fish. They are very ugly looking, and through the constant eating of alligators their faces and bodies are full of big lumps and growth which makes them look like demons. Their morals are so low that it is impossible to write them. In short they are a bunch of filthy people. Having convinced myself of their habits I did not go near them again but there sure is a difference between the Moros and the Tupí Indians, all living in the same landing and surroundings.

For awhile I made up my mind to let go of the rubber and stay the rest of my life around the lake, but being all alone I got despondent after which I would work and work with hope of getting back to civilization again, away from the pestering mosquitoes and pium (black flies). Also having a decent meal again. My hair had grown down to my back by now and I also had a beard tickling my bare chest, and no comb to comb out either my hair or my beard. I closed the ninth season with about four thousand pounds of rubber. After stowing it away with the rest, I went to the maloca to get the Indians to help me carry the canoe from the lake to the river, a distance of about three miles, so that I could make my way down the Boca do Xapurí, where I was to receive the rubber that I had given provision for.

Going down the river in an empty canoe was a gay trip, I made the trip in record time, eleven days. When I arrived at the Boca do Xapurí the people were in an uproar, for the Brazilian people had declared war on the Bolivians, because the Bolivians claimed that territory belonged to them and had erected a custom station at the mouth of the Acre river. So the first thing they done when I arrived was to ask me for my rubber. It was useless to tell them that I did not bring down any. My only worry was how to get the fellows that were to pay me back. I did not lose any time but went right back up the river to try to catch them in time. I had not much luck for the news spread and they came down the river without any rubber. Everybody left it behind for fear the men would take it from them. Going back to the station I waited for them and talked the matter over, with the results that if I wanted the rubber I would have to go and get it at the different places. I made up my mind right then not to mix in with that home made war, and went collecting my rubber, having gotten everything that belonged to me I tied it together and towed it up the river to hide it with the rest. A back-breaking undertaking to tow about four tons of rubber with a small canoe, forty days against the heavy current, but it was the only safe way out.

13

That trip finished me up. It took all the strength out of me. I was a very sick man. With nothing to eat but what I got hunting. I took a good rest with the Indians and then went back to tapping rubber. For awhile I did fairly well, say about two months. My feet then began to swell; at first I paid no attention but when I felt myself getting weaker and weaker I got alarmed, and gave up working. My appetite was gone. I went over to the maloca for help, but got so weak after awhile I had to ask them to bring me back to my camp. I had seen a good many die of the beri-beri but it never entered my mind that I was suffering from it. The Indians put me in a hangmat and stood some farinha which I had left over from the season before in my reach, they also put some water within my reach as everytime I got out of the hangmat I would drop fainting, and only with the greatest difficulty would I be able to get back into the hangmat again. The Indians showed themselves good friends. They would come around almost everyday making fire and giving me fresh water. I requested them to hide all my rubber in the grotto fundo which they did. The rainy season came and I was still laying there all swollen up and sleeping most of the time so that I very seldom saw the Indians when they came. They brought me some piquiva fruits which are very much like our plums. I knew then that it was around Christmas time. I do not know how long I laid there afterwards. When I did not show up in time at the station the boys surmised that there was something wrong. They all knew that I had gone back for the rubber without any provision so they reasoned that I had to come down this season for something, as it was impossible to hold over twice.

They waited until the steamer had unloaded and they then went looking for me. They did not know exactly where I was located, but here I learned what friendship meant. They located me with the help of the Tupís and brought me down to the Boca do Xapurí, only to find that the steamer had already left. The only thing to do now was to overtake the steamer to get me aboard. They loaded as much rubber as they could in the canoe and sent it down after the steamer. A dozen or more put me in a hangmat and put a pole through it and started off taking me over the land trying to head off the steamer. This is possible because the steamer very often takes on rubber until she reaches the mouth of the river Acre and then turns down the river Purus. Besides cutting off a good deal of the turns in the river, some of them take the steamer a day to turn, which if anyone knows the way can be made over land in an hour or two. They carried me seven days from day break until dusk, until they headed off the steamer. Always two men heading the way clearing the way so that they could get me through, committing an act of friendship I can never repay. They realized my condition and their idea was to get me away as far as possible down the river to change of water and air, which was the only chance in a condition of the beri-beri. The canoe reached the steamer and the boys put the rubber in the captain's care and I went down the river, but instead of getting better I got worse so the captain was forced to take me all the way down to Manaus calling a doctor on board. He had me transferred aboard another steamer that was leaving for Pará to get near the seashore. When we reached Pará I was completely helpless.

I was taken to a hotel and then a doctor was called who told me that I had an advanced state of beri-beri and was beyond help, and went away. After an hour or so he returned with another doctor and after giving me a thorough examination told me that there was a steamer leaving within an hour for Barbados, a British Island. They suggested that I take that steamer as it was the only hope. They offered to arrange my passage and to see me

A group of contemporary rubber gatherers, Rio Ituxí, Amazonas.

One of the portable growing boxes, Wardian cases, in
which rubber seedlings were transported from
England to Asia to break the Brazilian monopoly on
rubber and end the rubber boom.

aboard, so I was once more sent off. After six days of sailing we reached
the island and when the port authorities saw my condition they refused to
let me land. Of course I was helpless, but I never lost my senses and knew
everything that was going on. After much work the authorities seeing my
plight and knowing that I had sufficient funds they let me land. That is
they transferred me to the General Hospital where I was laying for eleven
months under the care of Doctor Graves until I was able to leave, taking
room at a hotel, under one arm was a crutch and a cane in the other hand.

My plan was to stay there until I was well enough that I could return to
the Amazonas and get my rubber and then go off to California. But Dr.
Graves advised me not to go to the Amazonas but to go to a colder climate
to get fresh blood. I then decided to return to the United States and remain
there until I would get strong enough to go back for my rubber. After
arriving at New York I put myself under the care of Dr. Burke who lives
between Twenty-ninth and Thirtieth St. for about four months. Also
taking Dr. Burke's advice not to return to the Amazonas I once more went
to work as a tile setter up to this date.

John C. Yungjohann

Epilogue

The following is a brief impression of my grandfather's life, after his return from the Amazonian region of Brazil in the early 1900's. After recuperating for eleven months in a Barbados hospital, John Christian Yungjohann, my maternal grandfather, returned to the United States.

In Barbados he had shared his hospital room with a baby monkey he had raised from the Amazonia, after its mother had died. The doctors allowed his companion to sit atop the bed post during his stay, along with an Amazon parrot, which had also accompanied them.

Often he expressed a desire to go back again to the Amazon region in an attempt to locate the *grotto fundo* (cave) in which he had stashed his most valuable crop of raw rubber during his final year there. However, his doctor advised him never to re-enter that tropical region, as it would surely send him into a relapse of the potentially deadly disease, beri-beri. His doctor believed the cold wet climate in New York best suited his condition. Perhaps ironically, the reverse might have been nearer to the truth.

Grandfather returned to his old profession, that of a tilesetter. Work he did during the construction of the Holland Tunnel symbolized the linking of his Dutch heritage to New World adventures. He was an adventurer as well as a fine artist. Tile mosaics completed in his spare time have found their way into the Museum of Art in New York City.

While living in Brooklyn, Jersey City, and Caldwell, New Jersey, my grandfather became a minister of the New Apostolic Church. It was said he held his congregation spellbound with tales of his Amazon adventures, all of which lend themselves vividly to the role faith plays in life's varied circumstances.

My grandfather had married Mary Louise Massa, who he had met at church. They had two children, my mother, Helen, and her brother, Edward. Mary Louise died during the New York flu epidemic when my mother was nine years old. A couple of years later my grandfather was married to Florence Berger, a women he adored until his death. This woman I knew as my grandmother and it was she who, at my never ending requests, would tell still another thrilling episode of my grandfather's Amazon jungle adventures.

John Christian Yungjohann died in 1930, following a lengthy illness. He had often commented on how difficult the winters were for him. One of his greatest pleasures while residing in New Jersey was spending warm summer days at a bungalow he built at the shore.

I've always felt a kinship with my grandfather that runs deeper than that of a grandson's obvious bloodline connection. The best way to describe this feeling is simply to say that my grandfather has always been very real to me, though he died thirteen years before I was born.

Now, when I visit my little bungalow each summer down in Key West, Florida and lie in my hammock gazing up at coconut palms and flowering frangipanis — somehow, I know he loves it here, as well.

Yungjohann   Hillman

Glossary

Acre The state of Brazil where the author worked as a rubber cutter. This territory formerly belonged to Bolivia and after a dispute was ceded to Brazil in 1903 in exchange for an indemnity and a promise to build the Madeira-Mamoré railroad which gave Bolivia access to shipping at the Brazilian river port of Porto Velho. The text refers to a later small dispute in the region.

araba A local measure of weight for a basket of 60 kilos.

barraca The Portuguese for a tent or hut. Used here for the shelters constructed by the author.

beijú A tapioca cake made from the cassava root.

Boca do Xapurí The mouth of the Xapurí river. Many settlements at river mouths are so named in Amazonia. This is at the junction of the Rio Xapurí and Rio Acre where the author returned each year to sell rubber and buy supplies.

boracha The Portuguese for rubber.

cachaça The local sugar cane liquor, popular throughout Brazil. It is a white rum.

calabash A gourd made from the fruit of the calabash tree, *Crescentia cujete,* which is the most used utensil in Amazonia. It serves as a water pot, soup bowl, canoe bailer, etc.

capybara The world's largest rodent which is a native of tropical America. This web-footed, pig-sized animal can weigh up to 100 pounds and is good eating.

carne seca Dried meat.

caucho The rubber latex produced from the *Castilla elastica* tree in the Fig Family. It is also called caoutchouc, hence the writer's word kautxuck.

estrada Literally a road. In the sense used here it is the rubber cutters trail through the forest linking a series of rubber trees.

facão A machete.

farinha The coarse flour produced from the cassava or manioc root. It must be processed properly to remove cyanide poison which occurs in the unprocessed root.

fazenda A farm.

jacaré The word for all species of caiman, the Amazon alligators.

kautxuck (See caucho)

maloca The communal dwelling house of Indian tribes.

mandioca root The cassava plant, *Manihot esculenta* which is the staple of Amazonian residents. (See also beijú and farinha.)

mateiro A forest guide or woodsman. Used here for the person who opened up the rubber paths.

mil-réis The former Brazilian unit of currency. At the time this account was written there were three mil-réis to the dollar.

mixire A fermented beer made from the cassava root. The fermentation is started by human saliva. The Indians chew bits of the root and put them in the liquid.

Pará A large state of Brazil at the mouth of the Amazon. The capital is Belém.

piranha A flesh eating fish with an exaggerated reputation for ferocity. These fish are attracted by blood and can be dangerous, but it is safe to swim in most Amazon waters without danger of being attacked.

pium Small black flies of the Family Simulidae. They occur in great numbers in the basins of the Purus and Madeira rivers.

queixada The white-lipped pecary, one of the wild pigs of Amazonia.

Rio Xapurí The river where the author spent his ten years rubber cutting. This river is a tributary of the Acre river which in turn flows into the Purus.

seringa The Portuguese for rubber latex.

seringal A rubber cutters area of rubber trees or a rubber tree plantation.

seringueira A rubber tree, *Hevea brasiliensis* of the Spurge Family, Euphorbiaceae.

seringueiro A rubber gatherer.

tabatinga A sedimentary type of clay used for pottery.

Tupí Indians One of the major language groups of Amazonian Indians.

Further Reading about the Amazon Rubber Boom

Boxer, C.R., *The Golden Age of Brazil 1659-1750: Growing Pains of a Colonial Society*. University of California Press, Berkeley, California,1975.

Collier, R., *The River That God Forgot: The Story of the Amazon Rubber Boom*. E.P. Dutton, New York, 1968.

Ferreira de Castro, J.M., *The Jungle: A Tale of the Amazon Rubber Boom*. E.P. Dutton, New York, 1934.

Hemming, J., *Red Gold: The Conquest of the Brazilian Indians*. Harvard University Press, Cambridge, Massachusetts, 1978.

Macedo, Soares, J.C., *A Borracha: Estudo Economico e Estatistiro*. Paris, 1927.

Pearson, H.C., *The Rubber Country of the Amazon*. India Rubber World, New York, 1911.

Rivera, J.E. [trans. by E.K. James], *The Vortex (La Voragine)*. G.P. Putnam & Sons, New York, 1935.

Weinstein, B., *The Amazon Rubber Boom 1850-1920*. Stanford University Press, Stanford, California, 1983.

Wolf, H. & Wolf, R., *Rubber: A Story of Glory and Greed*. Covici, Friede, New York, 1936.

Notable Works in Biospherics and Cultural Anthropology

The Biosphere Catalogue
Edited by Tango Parrish Snyder ($12.95)

The Biosphere Catalogue pulls together state of the art information by leading scientists and thinkers in the various spheres of the biosphere — biomes, atmosphere, geosphere, evolution, communication, analytics, cities, travel, and more.

Space Biospheres
By John Allen and Mark Nelson ($6.95)

Commencing with a brilliant and concise integrative model of Biosphere I (Earth's biosphere), the authors go on to present the modelling of Biosphere II, an extraordinary project in the creation of biospheric systems to further humankind's ability to live in harmony with the sphere of live — on Earth or among the stars.

The Biosphere
By Vladimir Vernadsky ($5.95)

The first English publication of the original theory of the biosphere written in 1926. Preface by Dr. Evgenii Shepelev from the Institute of Biomedical Problems, Moscow.

Traces of Bygone Biospheres
By Andrey Lapo ($9.95)

The role of life in geological processes is revealed via the study of biomineralization and biosedimentation phenomenon — the traces of bygone biospheres.

Where The Gods Reign:
Plants and Peoples of the Colombian Amazon
By Richard Evans Schultes ($20.00)

Journey the mysterious Colombian Amazon with this most remarkable scientist and eloquent guide. With 140 black and white photographs from 1940-1954, Dr. Schultes presents an exquisite and moving portrait of his historic studies of ethnobotany in this remote region.

The Renaissance of Tibetan Civilization
By Christoph von Führer-Haimendorf

Professor Haimendorf documents the extraordinary achievements of Tibetan refugees in reconstructing their civilization in India and Nepal since their flight from Tibet in 1959.

SP
Synergetic Press
Post Office 689   Oracle, Arizona 85623